SETS

WILLIAM W. FAIRCHILD
Assistant Professor of Mathematics

CASSIUS IONESCU TULCEA
Professor of Mathematics

Northwestern University

W. B. SAUNDERS COMPANY · PHILADELPHIA · LONDON · TORONTO

W. B. Saunders Company: West Washington Square
Philadelphia, Pa. 19105

12 Dyott Street
London, WC1A 1DB

1835 Yonge Street
Toronto 7, Ontario

Sets

SBN 0-7216-3540-7

Print No.:   9   8   7   6   5   4   3   2

PREFACE

The notions of set and function are introduced here and are discussed in some detail.

Although we make several remarks concerning various axioms of set theory, we do not *explicitly* base our presentation on a set of axioms. While the reader should be warned about the non-validity of certain statements that may seem intuitive (for instance, there is no set containing as its elements *all* the sets), it is difficult, at this level, to present set theory axiomatically.* A thorough presentation of axiomatic set theory requires a description of formal systems.

The main purpose of this text is to provide a basis for the study of various other fields of mathematics. The material we present is divided into thirteen chapters and two appendices.

In Chapter 1 we introduce the notion of a set. After clarifying its meaning by a series of examples, we describe the use of the symbols "$\in$" and "$\subset$."

In Chapter 2 we define the union, the intersection and the difference of two sets. We also define the complement of one set with respect to another. We establish various properties, for instance, the distributivity of intersection with respect to union and the duality (or de Morgan) formulas.

In Chapter 3 we introduce the Cartesian product of two sets and the Cartesian product of n ($n > 1$) sets.

In Chapter 4 we introduce the notion of function. A function is essentially defined as a subset of a Cartesian product, having certain properties. A long series of examples of functions is given.

---

* The interested reader may consult [1], [2], [7] and [10].

We also show how to represent by a table a function having finite domain.

In Chapter 5 we introduce the image and the inverse image of a set by a function, and we establish a series of results that will have useful applications.

Chapters 6 and 7 deal with the composition of two functions and with the notion of inverse function.

Chapter 8 deals with families, and in particular with families of sets. We define the union, the intersection and the Cartesian product of a family of sets and we establish various properties.

In Chapter 9 we define equipotent sets. We give a series of examples and we establish the Schroeder-Bernstein theorem.

Chapter 10 deals with relations. We discuss in some detail the notion of equivalence relation, since it is very often used in mathematics.

In Chapter 11 we introduce and discuss order relations and state Zorn's lemma, which has many applications in mathematics. Cardinal numbers are discussed briefly.

Chapter 12 deals with mathematical induction in the setting of countable sets.

Chapter 13 is concerned with combinatorial analysis. We establish here, as an application of the notions that were previously introduced, various elementary formulas.

In Appendix I we discuss some of the properties of real numbers that we use in this text.

In Appendix II we introduce the signature of a permutation. Although it is not necessary that the proofs in this Appendix be studied on a first reading, it is recommended that the statements of Theorem II.10, Corollary II.12, and Proposition II.6 be retained. The content of this Appendix (as well as some of the formulas in Chapter 13) is useful in multilinear algebra, which in turn is basic, for instance, in the study of multidimensional calculus.

At the end of each chapter there are exercises, the solving of which will yield additional insight to the material. It will be very useful if the student tries to formulate and solve related problems.

The authors wish to express their thanks to the Editorial staff of the Saunders Company for their cooperation and especially to Mr. George Fleming for his helpful suggestions concerning both the form and content of the manuscript. They also wish to thank Mrs. Pamela Frye and Mrs. Mae Leeds for their faithful and accurate typing of the manuscript.

SOME REMARKS
ABOUT NOTATION

1. Passages appearing between the triangles ▼ and ▲ may be omitted on a first reading.

2. The symbol **Z** is sometimes used to indicate a statement the understanding of which may cause difficulty.

3. The more difficult exercises are denoted by an asterisk.

The Greek Alphabet

The Greek alphabet is listed below since some of its letters are used in the text.

A	α	alpha	N	ν	nu
B	β	beta	Ξ	ξ	xi
Γ	γ	gamma	O	o	omicron
Δ	δ	delta	Π	π	pi
E	ε	epsilon	P	ρ	rho
Z	ζ	zeta	Σ	σ	sigma
H	η	eta	T	τ	tau
Θ	θ	theta	Υ	υ	upsilon
I	ι	iota	Φ	φ	phi
K	κ	kappa	X	χ	chi
Λ	λ	lambda	Ψ	ψ	psi
M	μ	mu	Ω	ω	omega

CONTENTS

Chapter I

Sets

In this book we are basically concerned with the notion of set. We shall consider this notion to be primitive, and shall clarify its meaning by several examples.

(1) Consider the set of all books in a certain bookstore. Each book in the store is an object of this set.

(2) Consider the set consisting of the four letters u, v, w, z. The letter u is an object of this set; the letters v, w, z are also objects of this set.

(3) The set consisting of the integers $1, 2, 3, \ldots, n, \ldots$, will be denoted N. The integers in this set are called *natural integers* or *natural numbers*.

(4) The set of all integers $\ldots, -3, -2, -1, 0, 1, 2, 3, \ldots$, will be denoted Z.

(5) The set of all rational numbers will be denoted Q.

(6) The set of all real numbers will be denoted R.

We assume in this text that the sets N, Z, Q, and R and their usual properties are known. For more discussion and a bibliography on the subject, see Appendix I.

It is often convenient to exhibit a set by displaying its objects between brackets. Thus, the set we gave in Example (2) can be written $\{u, v, w, z\}$.* Concerning this notation, we note that the order in which the objects are written is not relevant. For instance, $\{u, v, w, z\}$ and $\{u, w, z, v\}$ represent the same set.

---

* Whenever we use this notation, we agree that we may list the same element more than once. For instance, $\{1, 1, 2, 3, 3\}$ is the set whose elements are 1, 2, and 3; hence $\{1, 1, 2, 3, 3\} = \{1, 2, 3\}$.

(7) The set $\{a, -1, 4, 7, d, 0\}$ consists of the letters a and d and the numbers -1, 4, 7, and 0. Note that

$$\{a, -1, 4, 7, d, 0\} = \{a, d, -1, 4, 7, 0\}.$$

If t is an object, then we denote by $\{t\}$ the set consisting of the single object t. We distinguish, therefore, between the *object t* and the *set* $\{t\}$. We will give a justification for this practice later. A helpful non-mathematical example was suggested by P. R. Halmos: A box containing a hat is not the same thing as a hat.

Besides having sets consisting of *one* object, we shall consider a set that "does not contain any object." This set will be denoted $\varnothing$ and is called the *void set* or the *empty set*. It will be seen later how convenient this set is.

We shall now introduce and discuss certain notations.

We shall usually denote sets by capital roman letters; objects (at least in theoretical discussions) will often be denoted by small roman letters. Instead of *object* we shall sometimes say *element*. For instance, in Example (2), given earlier, u is an element of the considered set.

Let now A be a set and a an object. The notation

1.1 $a \in A$

means that a is an object of A. The notation 1.1 is read, "a is an element of A," or "a is in A," or even "a in A."

The notations $a \in A$ and $A \ni a$ are considered to be equivalent. The notation $A \ni a$ is read "A contains a (as an element)."

If we consider the set $\{u, v, w, z\}$, we have, for instance, $u \in \{u, v, w, z\}$. If we consider the set $\mathbf{Z}$, for instance, then $-1 \in \mathbf{Z}$.

The notation

1.2 $a \notin A$

means that a *is not an object of* A. The notation 1.2 is read, "a is not an object of A," or "a is not in A," or "a not in A."

The notations $a \notin A$ and $A \not\ni a$ are considered to be equivalent. The notation $A \not\ni a$ is read, "A does not contain a."

Example 1.—We have $\frac{1}{3} \notin \mathbf{Z}$, $-1 \in \mathbf{Q}$, $\mathbf{Z} \not\ni \frac{1}{5}$, $\mathbf{Z} \ni 5$, $\mathbf{Q} \ni \frac{1}{5}$. We have $\sqrt{2} \in \mathbf{R}$ and $\sqrt{2} \notin \mathbf{Q}$ (see Appendix I). If $B = \{1, 2, 3, 4, 5, 6, 7\}$, then $x \in B$ if and only if x is one of the numbers 1, 2, 3, 4, 5, 6, 7.

Note also that $x \in \{t\}$ if and only if $x = t$. Also, $y \notin \varnothing$ (or equivalently, $\varnothing \not\ni y$) for *any* object y.

Let now A and B be sets. The notation

1.3 $A \subset B$

means that

$$x \in A \text{ implies } x \in B.*$$

Hence, $A \subset B$ if and only if every element of A is an element of B; there can be, of course, elements of B that are not elements of A. The notation 1.3 is read, "A is contained in B." If $A \subset B$, then we say that A is a *part* or a *subset* of B.

The notations $A \subset B$ and $B \supset A$ are considered to be equivalent. The notation $B \supset A$ is read, "B contains A."

Remarks.—(i) Let A be a set and a an object. Then $a \in A$ if and only if $\{a\} \subset A$.

(ii) If a is any object, we have $a \in \{a\}$; we do *not* have $a \subset \{a\}$.
(iii) For every set A, we have $A \supset \varnothing$ and $A \supset A$.

Example 2.—We have $\{1, 2, 3, 4\} \subset \{1, 2, a, 3, 4\}$. We have $\varnothing \in \{\varnothing, \{\varnothing\}\}$.

Let again A and B be two sets. The notation

1.4 $A \not\subset B$

means that A *is not contained in* B. This, of course, means that there exists an object that belongs to A and that does not belong to B. The notation 1.4 is read, "A not contained in B."

The notations $A \not\subset B$ and $B \not\supset A$ are considered to be equivalent. The notation $B \not\supset A$ is read, "B does not contain A."

Example 3.—We have $N \subset Q$. However, $Q \not\subset N$ since, for instance, $\frac{1}{5} \in Q$ but $\frac{1}{5} \notin N$. We have $Q \subset R$ and $Q \neq R$ (notice that $\sqrt{2} \in R$ and $\sqrt{2} \notin Q$).

Two sets, A and B, are *equal* if they consist of the same objects. Therefore, they are equal if $x \in A$ *implies* $x \in B$ and $x \in B$ *implies*

---

* A statement such as "$x \in A$ implies $x \in B$" is equivalent to "for every x, $x \in A$ implies $x \in B$."

$x \in A$. Hençe:

1.5 $A = B$ *if and only if $A \subset B$ and $B \subset A$.*

Before proceeding further, we state the following properties:*

1.6   Let A, B, and C be three sets. Then
  (i) $A \subset A$;
  (ii) $A = B$ if and only if $B \subset A$ and $A \subset B$;
  (iii) $A \subset B$ and $B \subset C$ implies $A \subset C$.

For every set X, we denote by $\mathscr{P}(X)$ *the set of all subsets of X.*
Note that the elements of $\mathscr{P}(X)$ are also sets (namely, sets that are
subsets of X). Note that:

$$\varnothing \in \mathscr{P}(X) \quad \text{and} \quad X \in \mathscr{P}(X).$$

Example 4.—(i) If $X = \{1, 2, 3\}$, then $\mathscr{P}(X) = \{\varnothing, \{1\}, \{2\},$
$\{3\}, \{2, 3\}, \{1, 3\}, \{1, 2\}, \{1, 2, 3\}\}$.
  (ii) If $Y = \{a\}$, then $\mathscr{P}(Y) = \{\varnothing, \{a\}\}$.
  (iii) If $Z = \varnothing$, then $\mathscr{P}(Z) = \{\varnothing\}$. Note that $\mathscr{P}(Z) \neq \varnothing$.

Sets of the form $\mathscr{P}(X)$ will often be used in what follows.
We shall now explain three symbols that will frequently be
used to abbreviate certain statements. The symbols are given on the
left of the page; on the right we indicate the word or words that
these symbols may replace.

$\Rightarrow$ *implies*
$\Leftarrow$ *is implied by*
$\Leftrightarrow$ *if and only if* or *is equivalent to*

Example 5.—Let A, B, and C be three sets. Then

$$A \subset B \quad \text{and} \quad B \subset C \Rightarrow A \subset C;$$
$$a \in A \quad \text{and} \quad A \subset B \Rightarrow a \in B;$$
$$A = B \Leftrightarrow A \subset B \quad \text{and} \quad B \subset A;$$
$$A \subset B \Leftrightarrow (x \in A \Rightarrow x \in B);$$
$$t \text{ is an object} \Rightarrow t \notin \varnothing;$$
$$a \notin B \Leftrightarrow B \not\ni a.$$

  (8) We denote by $\mathbf{Z}_+$ the set of all positive integers.
  (9) We denote by $\mathbf{Q}_+$ the set of all positive rationals.
  (10) We denote by $\mathbf{R}_+$ the set of all positive real numbers.

---

* The first two have been already discussed above. The proof of the third is obvious
and is left to the reader.

In our terminology, the number 0 is positive. A number (integer, rational, or real) is *strictly positive* if it is positive and $\neq 0$. A number is *negative* if it is not strictly positive. A number is *strictly negative* if it is negative and $\neq 0$.

The positive integers are the numbers $0, 1, 2, \ldots$. The strictly positive integers are the numbers $1, 2, \ldots$. A rational number is positive if and only if it can be written in the form n/m with $n \in Z_{+}$, $m \in N$. A rational number is strictly positive if and only if it can be written in the form n/m with $n \in N$, $m \in N$.

Certain sets can be conveniently written by using the following notation. After several examples, the reader should understand how to use this notation.

   (i) The notation $\{x \mid x \in N, x \geq 4\}$ represents the set consisting of the integers $4, 5, 6, 7, \ldots$.

   (ii) The notation $\{x \mid x \in N, 0 \leq x \leq 5\}$ represents the set $\{0, 1, 2, 3, 4, 5\}$.

   (iii) $\{x \mid x \in Z, x^2 = 4\} = \{-2, +2\}$.

   (iv) $\{x \mid x \in Q, x^2 = 2\} = \varnothing$ (see Appendix I).

   (v) $\{x \mid x \in R, x^2 = 2\} = \{-\sqrt{2}, \sqrt{2}\}$.

   (vi) $\{x \mid x \in Z, x \geq 1\} = N$.

   (vii) If X is any set, then $\{A \mid A \subset X\} = \mathscr{P}(X)$.

The set $\{x \mid x \in N, x \geq 4\}$ is read, "the set of all $x \in N$ such that $x \geq 4$." The set $\{x \mid x \in Q, x^2 = 2\}$ is read, "the set of all $x \in Q$ such that $x^2 = 2$." In a similar way, we read the other sets just given.

Exercises for Chapter 1

1. Decide whether the following assertions are true or false:

   (a) $\{1, a, 2, 3\} \subset \{1, 2, 3, a\}$;
   (b) $\{1, a\} \in \{1, a, 2\}$;
   (c) $\{y, x\} \subset \{x, y\}$;
   (d) $\{x, y\} \subset \{\{x\}, \{x, y\}\}$;
   (e) $x \in \{\{x\}, \{x, y\}\}$.

2. Decide whether the following assertions are true or false:

   (a) $\varnothing \in \varnothing$;
   (b) $\varnothing \subset \varnothing$;
   (c) $\varnothing \in \{\{\varnothing\}, x\}$;

(d) $\varnothing \in \{\varnothing\}$;

(e) $\{\varnothing, \{\varnothing\}\} \in \{\varnothing, \{\varnothing\}, \{\varnothing, \{\varnothing\}\}\}$;

(f) $\{\varnothing, \{\varnothing\}\} \subset \{\varnothing, \{\varnothing\}, \{\varnothing, \{\varnothing\}\}\}$;

(g) $\mathscr{P}(X) \subset \{\mathscr{P}(X)\}$ where X is a set.

3. Define $A_0 = \varnothing$, $A_1 = \{\varnothing\}$, $A_2 = \{\varnothing, \{\varnothing\}\}$, $A_3 = \{\varnothing, \{\varnothing\}, \{\varnothing, \{\varnothing\}\}\}$ and $A_4 = \{\varnothing, \{\varnothing\}, \{\varnothing, \{\varnothing\}\}, \{\varnothing, \{\varnothing\}, \{\varnothing, \{\varnothing\}\}\}\}$. Decide whether the following statements are true or false:

(a) $A_2 \in A_3$;

(b) $A_2 \subset A_3$;

(c) $A_1 \in A_3$;

(d) $A_1 \subset A_3$;

(e) $A_2 \in A_4$;

(f) $A_2 \subset A_4$.

4. Let X and Y be two sets. Show that

$$X \subset Y \iff \mathscr{P}(X) \subset \mathscr{P}(Y).$$

5. Construct the sets

(a) $\mathscr{P}(\{a, \{a, b\}\})$;

(b) $\mathscr{P}(\mathscr{P}(\mathscr{P}(\varnothing)))$.

6. Show that

$$\{\{x\}, \{x, y\}\} = \{\{a\}, \{a, b\}\} \iff x = a \quad \text{and} \quad y = b.$$

Chapter 2

Union, Intersection, Difference, Complement

We shall define and discuss in this section the union, the intersection, and the difference of two sets. We shall also define and discuss the complement of a set (with respect to a set containing the given set).

Union.—Let A and B be two sets. We define the set $A \cup B$ by

2.1 $$x \in A \cup B \Leftrightarrow x \in A \quad or \quad x \in B.$$

Hence, $A \cup B$ is the set of all objects that are *at least in one of the sets*, A or B.

The set $A \cup B$ is called the *union* of A and B; it is read, "A union B."

Example 1.—(i) $\{1, 4\} \cup \{2, 5\} = \{1, 2, 4, 5\}$.
(ii) $\{1, 3, 4\} \cup \{2, 3\} = \{1, 2, 3, 4\}$.
(iii) $\{1, 2\} \cup \{\{1, 2\}\} = \{1, 2, \{1, 2\}\}$.

Intersection.—Let A and B be two sets. We define the set $A \cap B$ by

2.2 $$x \in A \cap B \Leftrightarrow x \in A \quad and \quad x \in B.$$

7

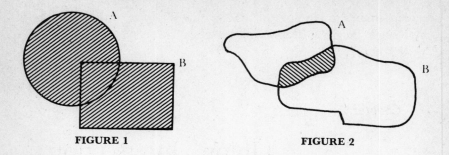

FIGURE 1   **FIGURE 2**

Hence, $A \cap B$ is the set of all objects that belong *to both sets*, A and B.

The set $A \cap B$ is called the *intersection* of A and B; it is read, "A intersection B."

Example 2.—(i) $\{1, 2, 3\} \cap \{1, 5\} = \{1\}$.

(ii) $\{1, 3\} \cap \{5, 7, 6\} = \varnothing$.

(Note that this intersection could not have been expressed without using the set $\varnothing$.)

(iii) $N \cap Z = N$.

(iv) $Z \cap Q = Z$; $Q \cap R = Q$.

The sets A and B are said to be *disjoint* if $A \cap B = \varnothing$. Hence, the sets in Example 2(ii) are disjoint. The sets $A_1, A_2, \ldots, A_p$ are said to be *pairwise disjoint* if $A_i \cap A_j = \varnothing$ for all $1 \leq i \leq n$, $1 \leq j \leq n, i \neq j$.

We shall now illustrate the union and intersection of two sets in the following way: Suppose A and B are as illustrated in Figure 1; then $A \cup B$ is the shaded part. Suppose now that A and B are as illustrated in Figure 2; then $A \cap B$ is the shaded part. If A and B are as in Figure 3, then $A \cap B = \varnothing$.

We shall give now some properties of the union and intersection of sets. We shall not prove here all the properties we state; however,

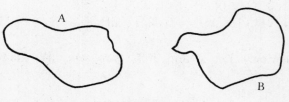

FIGURE 3

we recommend that the reader write the proofs of all the assertions (even when these assertions seem obvious).

2.3   *Let A be a set. Then*

   (i) $A \cup \varnothing = A, \quad A \cap \varnothing = \varnothing$;

   (ii) $A \cup A = A, \quad A \cap A = A.$

For example, by 1.5, $A \cup \varnothing = A$ if and only if $A \cup \varnothing \subset A$ and $A \subset A \cup \varnothing$. Let $x \in A \cup \varnothing$. Then $x \in A$ or $x \in \varnothing$. Since the set $\varnothing$ contains no element, we must have $x \in A$; therefore $A \cup \varnothing \subset A$. Conversely, if $x \in A$, then $x \in A$ or $x \in \varnothing$; hence $x \in A \cup \varnothing$; therefore $A \subset A \cup \varnothing$. This completes the proof.

2.4   *Let A and B be two sets. Then*

   (i) $A \cup B = B \cup A, \quad A \cap B = B \cap A$;

   (ii) $A \subset A \cup B, \quad A \supset A \cap B$;

   (iii) $A \supset B \Leftrightarrow A \cup B = A$;

   (iv) $A \supset B \Leftrightarrow A \cap B = B.$

2.5   *Let A, B, and C be three sets. Then*

   (i) $A \cup (B \cup C) = (A \cup B) \cup C$;

   (ii) $A \cap (B \cap C) = (A \cap B) \cap C.$

The assertions in 2.4(i) express the commutative property of the union and the intersection. Assertion 2.5(i) expresses the associative property of the union; assertion 2.5(ii) expresses the associative property of the intersection.

2.6   *Let A, B, A', and B' be four sets such that $A \subset A'$ and $B \subset B'$. Then:*

$$A \cup B \subset A' \cup B', \quad A \cap B \subset A' \cap B'.$$

It follows immediately from 2.6 and 2.3 that if A, B, and C are sets, $A \supset B$ and $A \supset C$, then $A \supset B \cap C$.

2.7   *Let A, B and C be three sets. Then*

$$A \cap (B \cup C) = (A \cap B) \cup (A \cap C).$$

Proof.—Let $X = A \cap (B \cup C)$ and $Y = (A \cap B) \cup (A \cap C)$. By 1.5 (Chap. 1), we have to show that $X \subset Y$ and $Y \subset X$.

We shall show first that $X \subset Y$. Let $x \in X$; then $x \in A$ and $x \in B \cup C$. Since $x \in B \cup C$, it follows that $x \in B$ or $x \in C$. If $x \in B$, then $x \in A \cap B$, whence

$$x \in (A \cap B) \cup (A \cap C);$$

if $x \in C$, then $x \in A \cap C$, whence

$$x \in (A \cap B) \cup (A \cap C).$$

Hence, $x \in X \Rightarrow x \in Y$.

We shall now show that $Y \subset X$. For this, we reason as follows. From 2.4 and 2.6 we know that :

$$B \cup C \supset B; \quad \text{hence} \quad A \cap (B \cup C) \supset A \cap B;$$

$$B \cup C \supset C; \quad \text{hence} \quad A \cap (B \cup C) \supset A \cap C.$$

By 2.6 and by 2.3

$$X = A \cap (B \cup C) \supset (A \cap B) \cup (A \cap C) = Y,$$

thus, $Y \subset X$.

Since $X \subset Y$ and $Y \subset X$, we conclude $X = Y$.

2.8   *Let A, B, and C be three sets. Then*

$$A \cup (B \cap C) = (A \cup B) \cap (A \cup C).$$

Proof.—Let $X = A \cup (B \cap C)$ and $Y = (A \cup B) \cap (A \cup C)$. By 1.5 (Chap. 1), we have to show that $X \subset Y$ and $Y \subset X$.

We shall show first that $X \subset Y$. We have $A \subset A \cup B$ and $A \subset A \cup C$, whence

$$A \subset (A \cup B) \cap (A \cup C).$$

Also, $B \cap C \subset B \subset A \cup B$ and $B \cap C \subset C \subset A \cup C$; hence

$$B \cap C \subset (A \cup B) \cap (A \cup C).$$

We deduce

$$X = A \cup (B \cap C) \subset (A \cup B) \cap (A \cup C) = Y,$$

hence, $X \subset Y$.

We shall show now that $Y \subset X$. Let $y \in Y$; then $y \in A \cup B$ and $y \in A \cup C$. If $y \in A$, then $y \in A \cup (B \cap C)$; that is, $y \in X$. Suppose now that $y \notin A$. Since $y \in A \cup B$, it follows that $y \in B$; since $y \in A \cup C$, it follows that $y \in C$. Hence, $y \in B \cap C$, and hence $y \in A \cup (B \cap C)$; that is, $y \in X$. Therefore, $Y \subset X$.

Since $X \subset Y$ and $X \supset Y$, we conclude $X = Y$.

The assertion in 2.7 expresses that the intersection is distributive with respect to the union; assertion 2.8 expresses that the union is distributive with respect to the intersection.

Difference.—Let A and B be two sets. We define the set $A - B$ by:

2.9 $x \in A - B \Leftrightarrow x \in A$ *and* $x \notin B$.

Hence, $A - B$ is the set of all objects that belong to A but do not belong to B. The set $A - B$ is called the *difference* of A and B; it is read, "A minus B."

Example 3.—(i) $\{1, 2, 5\} - \{1, 2\} = \{5\}$.
(ii) $\{1, 2, 3\} - \{2, 4\} = \{1, 3\}$.

Note that for every set A and B, we have $A - B \subset A$. Also, for every set A:

$$A - \varnothing = A, \qquad A - A = \varnothing.$$

If A is a set and $a \in A$, then $A - \{a\}$ is the set of all elements $x \in A$ that are distinct from a.

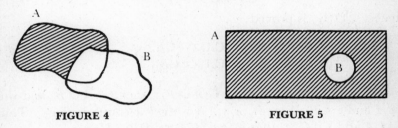

FIGURE 4   **FIGURE 5**

If A and B are as illustrated in Figure 4, then $A - B$ is the shaded part. Likewise, if A and B are as illustrated in Figure 5, then $A - B$ is the shaded part.

Complement.—Let X be a set and let $A \subset X$. We define the set $\mathbf{C}_X A$ by:

2.10 $\mathbf{C}_X A = X - A$.

The set $\mathbf{C}_X A$ is called the *complement of A with respect to X*. When there is no ambiguity as to what set X is, we write $\mathbf{C} A$ instead of $\mathbf{C}_X A$ and we call $\mathbf{C} A$ the *complement* of A. The notations $\mathbf{C}_X A$ and $\mathbf{C} A$ are read, respectively, "complement of A with respect to X" and "complement of A."

We shall now give some properties concerning the complement of a set. In 2.11 *and* 2.12 *we consider a fixed set X and subsets A and B of X. The complements are complements with respect to X.

2.11   (i) $A \cap \mathbf{C}A = \varnothing$, $A \cup \mathbf{C}A = X$.
(ii) $\mathbf{C}(\mathbf{C}A) = A$.
(iii) $\mathbf{C}\varnothing = X$, $\mathbf{C}X = \varnothing$.
(iv) $A \subset B \Leftrightarrow \mathbf{C}A \supset \mathbf{C}B$.

The proofs of (i), (ii), and (iii) are left to the reader.

Proof of 2.11(iv).—Suppose $A \subset B$. Then

$$x \in \mathbf{C}B \Rightarrow x \notin B \Rightarrow x \notin A \ (\text{since } A \subset B) \Rightarrow x \in \mathbf{C}A.$$

Hence

$$A \subset B \Rightarrow \mathbf{C}B \subset \mathbf{C}A.$$

Suppose now $\mathbf{C}B \subset \mathbf{C}A$. Using what we have already proved (note that $A \subset X$ and $B \subset X$ were arbitrary) and 2.11(ii), we obtain:

$$A = \mathbf{C}(\mathbf{C}A) \subset \mathbf{C}(\mathbf{C}B) = B.$$

Hence

$$\mathbf{C}B \subset \mathbf{C}A \Rightarrow A \subset B.$$

Hence, 2.11(iv) is proved.

2.12   (i) $\mathbf{C}(A \cup B) = (\mathbf{C}A) \cap (\mathbf{C}B)$.
(ii) $\mathbf{C}(A \cap B) = (\mathbf{C}A) \cup (\mathbf{C}B)$.

Proof of 2.12(i).—Let $x \in \mathbf{C}(A \cup B)$; then $x \notin A \cup B$, and thus $x \notin A$ and $x \notin B$ (if $x \in A$ or $x \in B$ we get a contradiction). Thus, $x \in \mathbf{C}A$ and $x \in \mathbf{C}B$; that is, $x \in (\mathbf{C}A) \cap (\mathbf{C}B)$. Hence

$$x \in \mathbf{C}(A \cup B) \Rightarrow x \in (\mathbf{C}A) \cap (\mathbf{C}B)$$

and hence

$$\mathbf{C}(A \cup B) \subset (\mathbf{C}A) \cap (\mathbf{C}B).$$

Now let $x \in (\mathbf{C}A) \cap (\mathbf{C}B)$; then $x \in \mathbf{C}A$ and $x \in \mathbf{C}B$, and thus $x \notin A$ and $x \notin B$. Thus, $x \notin A \cup B$, so $x \in \mathbf{C}(A \cup B)$. Hence

$$x \in (\mathbf{C}A) \cap (\mathbf{C}B) \Rightarrow x \in \mathbf{C}(A \cup B)$$

and hence

$$(\mathbf{C}A) \cap (\mathbf{C}B) \subset \mathbf{C}(A \cup B).$$

This completes the proof of 2.12(i).

Proof of 2.12(ii).—Let $A' = \mathbf{C}A$ and $B' = \mathbf{C}B$. Then, using 2.12(i), we deduce

$$(\mathbf{C}A) \cup (\mathbf{C}B) = A' \cup B' = \mathbf{C}(\mathbf{C}(A' \cup B'))$$
$$= \mathbf{C}((\mathbf{C}A') \cap (\mathbf{C}B')) = \mathbf{C}(A \cap B)$$

This completes the proof of 2.12(ii).

The results in 2.12(i) and 2.12(ii) (as well as certain generalizations that will be indicated later) are usually called the *duality formulas* or *De Morgan's formulas*.

▼   The properties given in 2.12 can be used to obtain new properties concerning sets from known ones. Suppose, for instance, that a certain relationship is true for any three sets. Then, it can be shown, using 2.12, that if we replace $\cup$ by $\cap$ and $\cap$ by $\cup$ we obtain a new relationship, which again is valid for any three sets. As an example concerning these remarks, note that, by 2.7, we have

$$A \cap (B \cup C) = (A \cap B) \cup (A \cap C)$$

for any three sets A, B, and C. If we replace $\cup$ by $\cap$ and $\cap$ by $\cup$, we obtain

$$A \cup (B \cap C) = (A \cup B) \cap (A \cup C)$$

that is, 2.8.   ▲

Let now $A_1, \ldots, \ldots, A_n$ be n $(n \in N)$ sets. In the same way as we defined $A \cup B$ and $A \cap B$, we may define the sets

$$A_1 \cup \ldots \cup A_n \quad \text{and} \quad A_1 \cap \ldots \cap A_n.$$

Thus, we define $A_1 \cup \ldots \cup A_n$ by: $x \in A_1 \cup \ldots \cup A_n$ if and only if there exists $i \in \{1, \ldots, n\}$ such that $x \in A_i$.

We define $A_1 \cap \ldots \cap A_n$ by: $x \in A_1 \cap \ldots \cap A_n$ if and only if $x \in A_j$ for every $j \in \{1, \ldots, n\}$.

The set $A_1 \cup \ldots \cup A_n$ is called the *union* of the sets $A_1, \ldots, A_n$; the set $A_1 \cap \ldots \cap A_n$ is called the *intersection* of the sets $A_1, \ldots, A_n$. The way in which these sets are read is obvious and will not be discussed here.

Exercise.—For any three sets A, B, C we have

$$A \cup B \cup C = A \cup (B \cup C) = (A \cup B) \cup C,$$
$$A \cap B \cap C = A \cap (B \cap C) = (A \cap B) \cap C.$$

Example 4.—(i) $\{1\} \cup \{2\} \cup \{3\} \cup \{4\} = \{1, 2, 3, 4\}$.
(ii) $\{1, 2, 3\} \cap \{0, 1, 2\} \cap \{2, 3, 4\} = \{2\}$.

Unions and intersections of arbitrary "families" of sets will be discussed in a later paragraph.

With the notation introduced at the end of Chapter 1, we can write
(i) $\{x \mid x \in A \quad \text{or} \quad x \in B\} = A \cup B$.
(ii) $\{x \mid x \in A \quad \text{and} \quad x \in B\} = A \cap B$.
(iii) $\{x \mid x \in A \quad \text{and} \quad x \notin B\} = A - B$.
(iv) $\{x \mid x \in X, x \notin A\} = \mathbf{C}_X A$ (here $A \subset X$).

Exercises for Chapter 2

1. Perform the indicated operations:
   (a) $\{1, 2, 3, 4\} \cup \{1, 5\} = ?$
   (b) $\{1, 2, 3\} \cup \mathbf{Z} = ?$
   (c) $\{a, b\} \cap \{1, 2\} = ?$
   (d) $(\{1, 2\} \cap \{1, 2, 4\}) \cup \{1, 2, 3\} = ?$
   (e) $\{1, 2\} \cap (\{1, 2, 4\} \cup \{1, 2, 3\}) = ?$
   (f) $\{\varnothing, \{1\}\} \cap \varnothing = ?$
   (g) $A_2 \cup A_3 = ?$ (See Exercise 3, Chap. 1.)
   (h) $A_2 \cap A_3 = ?$

2. Let A and B be two sets. Show that $A \cup B = A \cap B \Leftrightarrow A = B$.

3. Let A and B be two subsets of a set X. Show that

   (a) $A \cap \mathbf{C}B = \varnothing$ and $(\mathbf{C}A) \cap B = \varnothing \Leftrightarrow A = B$.
   (b) $((\mathbf{C}A) \cap (A \cup B)) \cup (A \cap B) = B$.

4. Let A, B, and C be three arbitrary subsets of a set X. Show that

$$(A \cap B) \cup C = A \cap (B \cup C) \Leftrightarrow C \subset A.$$

5. Give an example of a set X having two subsets, A and B, satisfying:

$$X - (A \cap B) \neq (X - A) \cap (X - B).$$

6. Let X be a set and suppose $\{E, F, G\} \subset \mathscr{P}(X)$. Show that

$$(E - G) \cap (F - G) = (E \cap F) - G.$$

7. Define, for each two sets A, B: $A \triangle B = (A - B) \cup (B - A)$.

   (a) Let $A = \{1, 3, 4\}$, $B = \{1, 5, 7\}$. Write out the set $A \triangle B$.
   (b) Write out the set $\mathscr{P}(A) \triangle \mathscr{P}(B)$ when A and B are as in 7(a).
   (c) Write out the sets $A_1 \triangle A_2$, $A_2 \triangle A_3$, and $A_1 \triangle A_3$ for which A_1, A_2, and A_3 are defined as in Exercise 3, Chapter 1.

Chapter 3

The Cartesian Product

Given two objects, x and y, we may form a new object, which we shall denote (x, y). We shall call this new object* a *couple*.

Two couples, (x', y') and (x'', y''), are identical if and only if

$$x' = x'' \quad \text{and} \quad y' = y''.$$

For example, $(x, y) = (y, x) \Leftrightarrow x = y$.

The reader should note that the notion of couple and that of set containing two elements are distinct; he should *not* write $(x, y) = \{x, y\}$.

Example 1.—(i) $(1, 2) \neq (2, 1)$.
  (ii) $(1, 2) = (\alpha, \beta) \Leftrightarrow 1 = \alpha$ and $2 = \beta$.
(iii) $\{1, 2\} = \{2, 1\}$.

Product.—Let A and B be two sets. We define $A \times B$ by

3.1 $\quad z \in A \times B \Leftrightarrow z = (x, y) \quad \text{with} \quad x \in A \quad \text{and} \quad y \in B.$

Hence, $A \times B$ is the set of all couples (x, y) such that $x \in A$ and $y \in B$.

The set $A \times B$ is called the *Cartesian product* of A and B or simply the *product* of A and B; it is read "A times B."

Note that if A and B are two sets, then $A \times B = \varnothing$ if and only if at least one of the sets, A or B, is void. Thus, $A \times \varnothing = \varnothing$ and $\varnothing \times A = \varnothing$ for every set A.

---

* Actually, the couple (x, y) may be defined to be the set $\{\{x\}, \{x, y\}\}$; then the property $(x', y') = (x'', y'') \Leftrightarrow x' = x''$ and $y' = y''$ follows from Exercise 6, Chapter 1.

Example 2.—(i) Let $A = \{1\}$ and $B = \{2\}$; then $A \times B = \{(1, 2)\}$.

(ii) Let $A = \{1, 2\}$ and $B = \{a, b, c\}$. Then

$$A \times B = \{(1, a), (2, a), (1, b), (2, b), (1, c), (2, c)\}.$$

Note that $A \times B$ has $6 (= 2 \times 3)$ elements if $a \neq b, b \neq c$, and $c \neq a$.

The product of sets is not "commutative." For instance, if A and B are as in Example 2(i), then $A \times B = \{(1, 2)\}$ and $B \times A = \{(2, 1)\}$; then $A \times B \neq B \times A$ because $(1, 2) \neq (2, 1)$.

Among the properties of the product of two sets, we shall state here the following:

3.2   (i) *Let A, B, A', B' be four sets such that $A \times B \neq \varnothing$. Then*
$$A \times B \subset A' \times B' \Leftrightarrow A \subset A' \text{ and } B \subset B'.$$
     (ii) *Let A, B, C be three sets. Then*
$$(A \cup B) \times C = (A \times C) \cup (B \times C).$$
     (iii) *Let A, B, A', B' be four sets. Then*
$$(A \times B) \cap (A' \times B') = (A \cap A') \times (B \cap B').$$
     (iv) *Let A, B, C be three sets. Then*
$$(A - B) \times C = A \times C - B \times C.$$

As an example, we shall prove 3.2(ii); we leave the proofs of the other three assertions to the reader.

Proof of 3.2(ii).—Let $L = (A \cup B) \times C$ and $K = (A \times C) \cup (B \times C)$. By 1.5 (Chap. 1), we have to show that $L \subset K$ and $K \subset L$.

We shall show first that $L \subset K$. Let $z \in L$. Then $z = (x, y)$ with $x \in A \cup B$ and $y \in C$. Since $x \in A \cup B$ we have either $x \in A$ or $x \in B$, whence $(x, y) \in A \times C$ or $(x, y) \in B \times C$. We deduce

$$z = (x, y) \in (A \times C) \cup (B \times C),$$

that is, $L \subset K$.

We shall now show that $K \subset L$. Let $z \in K$. Then $z = (x, y)$ with $(x, y) \in A \times C$ or $(x, y) \in B \times C$. If $(x, y) \in A \times C$, then $x \in A$ and $y \in C$, whence $x \in A \cup B$ and $y \in C$; hence $(x, y) \in (A \cup B) \times C = L$. If $(x, y) \in B \times C$, then $x \in B$ and $y \in C$, whence $x \in A \cup B$ and $y \in C$; hence $(x, y) \in (A \cup B) \times C = L$. Thus $z \in K \Rightarrow z \in L$, that is, $K \subset L$.

Hence $L \subset K$ and $K \subset L$, whence $L = K$.

Exercise.—(i) Let $A, B, A',$ and B' be non-void sets. Then $A \times B = A' \times B'$ if and only if $A = A'$ and $B = B'$.

(ii) Let A, B, C be non-void sets. Then $(A \times B) \times C \neq A \times (B \times C)$.

The notion of couple can be extended as follows. If x_1, x_2, x_3 are three objects, we define a new object (x_1, x_2, x_3) by

$$(x_1, x_2, x_3) = ((x_1, x_2), x_3).$$

Such an object is called a *triple*. Two triples, (x', y', z') and (x'', y'', z''), are identical if and only if $(x', y') = (x'', y'')$ and $z' = z''$; hence, if and only if $x' = x''$, $y' = y''$ and $z' = z''$.

If x_1, x_2, x_3, and x_4 are four objects, we define a new object (x_1, x_2, x_3, x_4) by

$$(x_1, x_2, x_3, x_4) = ((x_1, x_2, x_3), x_4).$$

Such an object is called a *quadruple*. Two quadruples, (x', y', z', t') and (x'', y'', z'', t''), are identical if and only if $(x', y', z') = (x'', y'', z'')$ and $t' = t''$; hence, if and only if $x' = x''$, $y' = y''$, $z' = z''$ and $t' = t''$.

Example 3.—We have $(1, 2, 3, 4) \neq (2, 1, 3, 4)$. We have $(1, 1, 1, 1) = (a, b, c, d)$ if and only if $a = b = c = d = 1$.

Now let x_1, x_2, $\ldots$, x_n be $n(n > 1)$ objects. Supposing that* $(x_1, x_2, \ldots, x_{n-1})$ was already defined, we then define a new object $(x_1, x_2, \ldots, x_n)$ by

$$(x_1, x_2, \ldots, x_n) = ((x_1, x_2, \ldots, x_{n-1}), x_n).$$

Such an object is usually called an *n-tuple* (hence, a triple is a 3-tuple; a quadruple is a 4-tuple). As in the case of triples or quadruples, we show that two n-tuples $(x_1', x_2', \ldots, x_n')$ and $(x_1'', x_2'', \ldots, x_n'')$ are identical if and only if

$$x_1' = x_1'', x_2' = x_2'', \ldots, x_n' = x_n''.$$

Now let A_1, A_2, $\ldots$, A_n be n sets. In the same way as we defined $A \times B$, we may define the set $A_1 \times A_2 \times \ldots \times A_n$. Thus, we define $A_1 \times A_2 \times \ldots \times A_n$ by

$$z \in A_1 \times A_2 \times \ldots \times A_n$$

if and only if $z = (z_1, z_2, \ldots, z_n)$ with

$$z_1 \in A_1, z_2 \in A_2, \ldots, z_n \in A_n.$$

---

* For $n = 1$ we simply write x instead of (x).

The set $A_1 \times A_2 \times \ldots \times A_n$ is called the *Cartesian product*, or simply the *product*, of the sets $A_1, A_2, \ldots, A_n$.

Cartesian products of arbitrary "families" of sets will be discussed in a later section.

With the notation introduced at the end of Chapter 1, we can write

(1) $A \times B = \{(x, y) \mid x \in A \text{ and } y \in B\}.$

(2) $A_1 \times A_2 \times \ldots \times A_n = \{(x_1, x_2, \ldots, x_n) \mid x_j \in A_j \text{ for all } j = 1, 2, \ldots, n\}.$

If $A_1 = A_2 = \ldots = A_n = A$, we shall sometimes write A^n instead of $A_1 \times A_2 \times \ldots \times A_n$. Thus

$$A^1 = A, \qquad A^2 = A \times A, \qquad A^3 = A \times A \times A.$$

Exercises for Chapter 3

1. Let $A = \{1\}$ and $B = \{2, 3\}$. List the elements of $\mathscr{P}(A \times B)$. List the elements of $A \times \mathscr{P}(A \times B)$.

2. Perform the indicated operations:
   (a) $\{(1, 2), (2, 3), (3, 1)\} \cup \{(2, 2), 3\};$
   (b) $\{(2, 3)\} \cap \{(3, 2)\};$
   (c) $(\boldsymbol{N} \times \boldsymbol{Q}) \cap (\boldsymbol{Q} \times \boldsymbol{Z}).$

3. Let A, B, and C be three sets. Show that $A \times (B \cap C) = (A \times B) \cap (A \times C)$.

4. Find four sets, A, B, A', and B', such that

$$(A \times B) \cup (A' \times B') \neq (A \cup A') \times (B \cup B').$$

5.* Let X and Y be two sets. Any subset R of $X \times Y$ is said to be a *relation between X and Y*†; hence $\mathscr{P}(X \times Y)$ is the set of all relations between X and Y. Let Z be another set. If S is a relation between X and Y, and R is a relation between Y and Z, we denote by $R \circ S$ the set $\{(x, z) \mid x \in X, z \in Z, \text{ there is } y \in Y \text{ such that } (x, y) \in S, \text{ and } (y, z) \in R\}$. Clearly, $R \circ S$ is a relation between X and Z.

---

† If $X = Y$, we usually say that R is a relation in X, instead of saying that R is a relation between X and Y (see also Chapter 10). Notice that R is a relation between X and Y if and only if (X,Y,R) is a correspondence (see Exercise 18 in the section titled Supplementary Exercises).

Let $X = Y = Z = \{0, 1, 2, 3, 4\}$. Construct $R \circ S$ and $S \circ R$ in each of the following cases:

(a) $R = \{(0, 1), (1, 2)\}$, $S = \{(1, 2), (2, 3)\}$.
(b) $R = \{(0, 1), (1, 2)\}$, $S = \{(0, 1), (1, 2)\}$.
(c) $R = \{(0, 0), (1, 2)\}$, $S = \{(1, 2), (1, 0)\}$.
(d) $R = \{(0, 1), (0, 2), (0, 3)\}$, $S = \{(1, 0), (2, 0), (3, 0)\}$.

6.* Let W, X, Y, and Z be four sets, and suppose $T \in \mathscr{P}(W \times X)$, $S \in \mathscr{P}(X \times Y)$, $R \in \mathscr{P}(Y \times Z)$. Is it true that

$$(R \circ S) \circ T = R \circ (S \circ T)?$$

7.* Let X and Y be sets and suppose $R \subset X \times Y$.

(a) For each $A \subset X$, define $R[A]$ to be the set $\{y \mid \text{there is } x \in A \text{ such that } (x, y) \in R\}$. $R[A]$ is called the image of A by the relation R.
(b) For each $S \in \mathscr{P}(X \times Y)$, define $S^{-1} \in \mathscr{P}(Y \times X)$ to be the set $\{(y, x) \mid (x, y) \in S\}$.
(c) Show that for each $B \in \mathscr{P}(Y)$, $R^{-1}[B] = \{x \mid \text{there is } y \in B \text{ such that } (x, y) \in R\}$.

Chapter 4

Functions

We now introduce the notion of function. We shall give a series of examples.

4.1 Definition.—*A function is a triple $f = (A, B, G)$ such that:*

(i) *A, B, G are sets;*
(ii) *$G \subset A \times B$;*
(iii) *for each $x \in A$ there exists one and only one $y \in B$ such that $(x, y) \in G$.*

The reader should pay special attention to condition 4.1(iii).

The set A is called the *domain* of f; the set B is called the *range** of f. We usually write, $A = \operatorname{dom} f$ and $B = \operatorname{rng} f$. The set G is called the *graph†* of f; instead of G we shall sometimes write G_f.

Let $f = (A, B, G)$ be a function. If $x \in A$, then, by 4.1(iii), there exists *one and only one* $y \in B$ such that $(x, y) \in G$; this *unique y* will be denoted $f(x)$ and will be called the *value of f at x* (or the value of f for x). We note that

4.2 $$G = \{(x, f(x)) \mid x \in A\}.$$

Let f and g be two functions. Hence, f and g are triples:

$$f = (X', Y', G'), \qquad g = (X'', Y'', G'').$$

---

* Instead of *domain of f* we may say *departure set of f*; instead of *range of f* we may say *arrival set of f*.

† For the general notion of graph, the reader may consult the Supplementary Exercises.

Therefore (see Chap. 3), $f = g$ if and only if $X' = X''$, $Y' = Y''$, and $G' = G''$; thus *two functions are equal if and only if they have the same domain, the same range, and the same graph.*

We denote by

4.3 B^A or $\mathscr{F}(A, B)$ the set of all functions having A for domain and B for range.

Let f and g be two elements of $\mathscr{F}(A, B)$. In this case, $f = g$ if and only if $G_f = G_g$. Using 4.2, we deduce that $f = g$ if and only if

4.4 $\qquad\qquad f(x) = g(x) \quad$ for all $\quad x \in A$.

Note that to *define* a function f, it is enough (and necessary) to define its *domain*, its *range*, and for each $x \in \operatorname{dom} f$, the element $f(x) \in \operatorname{rng} f$.

Example 1.—Let $A = \{1\}$, $B = \{1, 2\}$ and $G = A \times B$. Then (A, B, G) is not a function. In fact, $(1, 1) \in G$ and $(1, 2) \in G$; hence, there is more than one $y \in B$ such that $(1, y) \in G$. Hence, 4.1(iii) is not satisfied, and hence (A, B, G) is not a function.

Example 2.—Let $A = \{0, 1, 2, 3\}$, $B = \{a, b, c\}$ (we assume $a \neq b$, $b \neq c$ and $c \neq a$), and $G = \{(0, a), (1, a), (2, a), (3, c)\}$. Then (A, B, G) is a function. If we write $f = (A, B, G)$, then $f(0) = a, f(1) = a, f(2) = a, f(3) = c$.

Example 3.—Let $G = \{(x, x^2) \mid x \in \mathbf{R}\}$. Then the triple $g = (\mathbf{R}, \mathbf{R}, G)$ is a function. For each $x \in \mathbf{R}$, we have $g(x) = x^2$.

Example 4.—For each $x \in \mathbf{R}$, let

$$|x| = x \qquad \text{if} \quad x \geq 0$$
$$= -x \quad \text{if} \quad x \leq 0.$$

(The number $|x|$ is called the absolute value of $x \in \mathbf{R}$.) Clearly

$$B = (\mathbf{R}, \mathbf{R}, \{(x, |x|) \mid x \in \mathbf{R}\})$$

is a *function.* We have $B(0) = 0$, $B(x) = |x|$, $B(-x) = B(x)$ for $x \in \mathbf{R}$.

Example 5.—Let X be a set. Then $f = (\varnothing, X, \varnothing)$ is a function, $\operatorname{dom} f = \varnothing$, $\operatorname{rng} f = X$, $G_f = \varnothing$.

Example 6.—Let A be a set and let $\Delta = \{(x, x) \mid x \in A\}$.* Then (A, A, Δ) is a function.

Properties 4.1(i) and 4.1(ii) are clearly satisfied. We shall now verify 4.1(iii). *Let* $x \in A$. Then $(x, x) \in \Delta$; thus, for each $x \in A$ there is $y \in A$ such that $(x, y) \in \Delta$. Suppose now that $(x, y') \in \Delta$ and $(x, y'') \in \Delta$. Since $(x, y') \in \Delta$, it follows that there is $t \in A$ such that $(x, y') = (t, t)$, whence $x = t$ and $y' = t$; that is, $x = y'$. In the same way, we see that $x = y''$. We deduce $y' = y''$. Hence, there is at most one $y \in A$ such that $(x, y) \in \Delta$, and hence 4.1(iii) is satisfied.

The function (A, A, Δ) is usually denoted by j_A; hence, we have $j_A(x) = x$ for all $x \in A$.

Example 7.—Let Y be a set, X a subset of Y, and $G = \{(x, x) \mid x \in X\}$. Then (X, Y, G) is a function.

Example 8.—Let W be a set and let $G = \{(A, \mathbf{C}A) \mid A \in \mathscr{P}(W)\}$. Then $\psi = (\mathscr{P}(W), \mathscr{P}(W), G)$ is a function and

$$\text{dom } \psi = \text{rng } \psi = \mathscr{P}(W).$$

For each $A \in \mathscr{P}(W)$, we have $\psi(A) = \mathbf{C}A$.

▼ *Example 9.*—Let Y be a set, and let $X = \mathscr{P}(W) \times \mathscr{P}(W)$ and $Y = \mathscr{P}(W)$. (Hence, an object belongs to X if and only if the object is of the form (A, B) with A and B subsets of W.) Now let

$$G_u = \{((A, B), A \cup B) \mid A \in \mathscr{P}(W), B \in \mathscr{P}(W)\};$$
$$G_i = \{((A, B), A \cap B) \mid A \in \mathscr{P}(W), B \in \mathscr{P}(W)\};$$
$$G_d = \{((A, B), A - B) \mid A \in \mathscr{P}(W), B \in \mathscr{P}(W)\}.$$

Then

$$f = (X, Y, G_u), g = (X, Y, G_i), h = (X, Y, G_d)$$

are functions. These three functions have the same domain, namely $\mathscr{P}(W) \times \mathscr{P}(W)$; they have also the same range, namely $\mathscr{P}(W)$.

For every $(A, B) \in \mathscr{P}(W) \times \mathscr{P}(W)$, we have

$$f((A, B)) = A \cup B, g((A, B)) = A \cap B, h((A, B)) = A - B \quad ▲$$

If f is a function, and if its domain consists of n-tuples ($n \in N$), then we shall usually write $f(x_1, x_2, \ldots, x_n)$ instead of $f((x_1, x_2, \ldots,$

———————————

* The set Δ is called the *diagonal* of A.

x_n)) for $(x_1, x_2, \ldots, x_n) \in \text{dom} f$. A function the domain of which consists of n-tuples is sometimes called a *function of n variables*.

▼   For instance, the elements in the domain of the functions f, g, h in Example 9 are couples (2-tuples). Hence, we may write

$$f(A, B) = A \cup B, \ g(A, B) = A \cap B, \ h(A, B) = A - B$$

for $(A, B) \in \mathscr{P}(W) \times \mathscr{P}(W)$. With this notation, we can express 2.7 and 2.8 (Chap. 2) as follows (here A, B, and C belong to $\mathscr{P}(W)$)

$$g(A, f(B, C)) = f(g(A, B), g(A, C)),$$

and

$$f(A, g(B, C)) = g(f(A, B), f(A, C)). \qquad ▲$$

Example 10.—Let A and B be two sets and let $A \times B$ be their product. Let

$$G_1 = \{((x, y), x) \mid (x, y) \in A \times B\}$$

and

$$G_2 = \{((x, y), y) \mid (x, y) \in A \times B\}.$$

Then $(A \times B, A, G_1)$ and $(A \times B, B, G_2)$ are functions. We usually denote $(A \times B, A, G_1)$ by pr_1 and call it *the projection* of $A \times B$ *onto* A. Note that

$$\text{dom} \ pr_1 = A \times B, \qquad \text{rng} \ pr_1 = A$$

and that

$$pr_1(x, y) = x \quad \text{for all} \quad (x, y) \in A \times B.$$

We usually denote $(A \times B, B, G_2)$ by pr_2 and call it the *projection* of $A \times B$ *onto* B.

Note that

$$\text{dom} \ pr_2 = A \times B, \qquad \text{rng} \ pr_2 = B$$

and that

$$pr_2(x, y) = y \quad \text{for all} \quad (x, y) \in A \times B.$$

Note also that, for every $z \in A \times B$, we have

$$z = (pr_1(z), pr_2(z)).$$

If $G \subset A \times B$ is such that 4.1(iii) is satisfied, then (A, B, G) is a function. Moreover, if $f = (A, B, G)$, then we have (see Chap. 5)

$$\text{dom} f = pr_1(G)$$

and

$$f(\text{dom} f) = pr_2(G).$$

Example 11.—Let $A = \{a_1, a_2, \ldots, a_p\}$ (we assume $a_i \neq a_j$ if $i \neq j$) and B be two sets, and let f be a function with domain $A \times A$ and range B. For every (i, j), where $1 \leq i \leq p$ and $1 \leq j \leq p$, let

$$b_{ij} = f(a_i, a_j).$$

The function f can be conveniently "represented by means of a table" as follows.

f	a_1	a_2		a_p
a_1	b_{11}	b_{12}		b_{1p}
a_2	b_{21}	b_{22}		b_{2p}
a_p	b_{p1}	b_{p2}		b_{pp}

An element in the column under f is to be considered as the *first* element of an ordered pair in the domain of f, while an element in the row to the right of f is to be considered as the *second* element in an ordered pair in the domain of f. If (a_i, a_j) is such a pair, then the value of f at (a_i, a_j) is the element of B that appears at the "intersection of the ith row and the jth column."

For instance, if $A = \{0, 1\}$ and $B = \{0, 1\}$, then the tables

M	0	1
0	0	1
1	1	1

and

m	0	1
0	0	0
1	0	1

represent respectively the functions

$$M = (A \times A, B, \{((0, 0), 0), ((0, 1), 1), ((1, 0), 1), ((1, 1), 1)\})$$

and

$$m = (A \times A, B, \{((0, 0), 0), ((0, 1), 0), ((1, 0), 0), ((1, 1), 1)\}).$$

A function $f = (X, Y, G_f)$ is said to be a *constant function* if there exists $\lambda \in Y$ such that*

$$f(x) = \lambda \quad \text{for all} \quad x \in X.$$

Note then that

$$G_f = \{(x, \lambda) \mid x \in X\}.$$

Conversely, any function the graph of which is of this form is clearly a constant function.

Whenever we say that f is a function *on X to Y*, we mean that f is a function with domain X and range Y. The notation

$$f : X \to Y$$

(which is sometimes read, "f X into Y") is very useful; it will always mean that f is a function with domain X and range Y. Occasionally, the notations

$$X \xrightarrow{f} Y \quad \text{or} \quad Y \xleftarrow{f} X$$

are used instead of $f : X \to Y$.

Instead of *function*, we shall sometimes say *mapping*. Thus, whenever we say that *f is a mapping of X into Y* (or *from X into Y*), we mean that f is a function with domain X and range Y.

Let f be a function with domain X and range Y. Sometimes we shall say that f is the mapping $x \mapsto f(x)$ of X into Y.

Whenever we say, "Consider the mapping $x \mapsto \theta(x)$ of X into Y,"† we mean that we consider the mapping

$$f = (X, Y, \{(x, \theta(x)) \mid x \in X\}).$$

Other notations than $\theta(x)$ can be used for $x \in X$. For instance, whenever we say, "Consider the mapping $x \mapsto x^3$ of R into R," we mean that we consider the mapping

$$(R, R, \{(x, x^3) \mid x \in R\}).$$

Whenever we say, "Consider the mapping $(x, y) \mapsto x + y$ of $R \times R$ into R," we mean that we consider the triple

$$(R \times R, R, \{((x, y), x + y) \mid (x, y) \in R \times R\}).$$

The reader should notice that we use two types of arrows, namely $\mapsto$ and $\to$. Whenever we write $f : X \to Y$, we mean that X

---

* Such a function is frequently denoted λ.
† Here $\theta(x) \in Y$ for all $x \in X$.

is the domain of f and Y is the range. The notation $f : x \mapsto \theta(x)$ or $x \mapsto \theta(x)$ means that x belongs to the domain of the considered function and $\theta(x)$ is the value of the function at x.

Exercises for Chapter 4

1. Is $\{(0, 1), (1, 2), (2, 3), (3, 1)\}$ the graph of some function?

2. Is $(\{0, 1\}, \{0\}, \{(0, 0), (1, 0)\})$ a function? Is $(\{0\}, \{0, 1\}, \{(0, 0,) (0, 1)\})$ a function?

3. Write down two distinct functions that have $\{(1, a), (2, a)\}$ as their graph.

4. Let $A = B = C = D = \{1, 2\}$. Is $(A \times B, C \times D, \{(2, 2, 1, 1)\})$ a function? Is $(A \times B, C \times D, \{(2, 0), (1, 2)\})$ a function? Is $(B \times C, A \times D, \{(2, 1), (2, 2)\})$ a function?

5. Let $X = \{a, b\}$ and denote by e_a the function $(\mathscr{F}(X, X), X, \{(f, f(a)) \mid f \in \mathscr{F}(X, X)\})$. List the elements of dom e_a and find the value of e_a at f for each $f \in$ dom e_a.

6. Construct the set $\{1, 2\}^{\{1\} \times \{1, 2\}}$.

7. Let $A = B = \{0, 1\}$. Does the table

C	0	1
0	1	1
1	0	1

"represent" (see Example 11) a function? If so, write the function as a triple.

Chapter 5

The Image and the Inverse Image of a Set by a Function

Let $f : X \to Y$ be a function. For every $A \subset X$, we write*

5.1 $$f(A) = \{f(x) \mid x \in A\}.$$

Hence, $f(A) \subset Y$ and an object y belongs to $f(A)$ if and only if there exists $x \in A$ such that $f(x) = y$.

The set $f(A)$ is called the *image* of A by f. Note that $f(\varnothing) = \varnothing$ and that $f(\{x\}) = \{f(x)\}$ for all $x \in X$.

Example 1.—Let f be the function in Example 2 of Chapter 4. Then

$$f(A) = \{a, c\}, \qquad f(\{0, 1\}) = \{a\}.$$

Note that $\operatorname{dom} f = A = \{0, 1, 2, 3\}$, $\operatorname{rng} f = B = \{a, b, c\}$ and that $f(A) = \{a, c\}$.

Example 2.—Let j_A be the function in Example 6 of Chapter 4. Then $j_A(D) = D$ for all $D \subset A$.

It is apparent from 5.1 that if $f = (X, Y, G)$ is a function, and if $D \subset C \subset X$, then

5.2 $$f(D) \subset f(C).$$

---

* This is an "abuse of notation" because the domain of f is X and not $\mathscr{P}(X)$.

Among other properties that are useful and easy to establish, we mention the following.

5.3 *Let $f = (X, Y, G)$ be a function and let $A \subset X$ and $B \subset X$ be two sets. Then:*

(i) $f(A \cup B) = f(A) \cup f(B)$;
(ii) $f(A \cap B) \subset f(A) \cap f(B)$;
(iii) $f(A) - f(B) \subset f(A - B)$.

Proof of 5.3(i).—Since $A \subset A \cup B$ and $B \subset A \cup B$, we have (using 5.2)

$$f(A \cup B) \supset f(A) \quad \text{and} \quad f(A \cup B) \supset f(B).$$

Hence

$$f(A \cup B) \supset f(A) \cup f(B).$$

Now let $y \in f(A \cup B)$. Then there is $x \in A \cup B$ such that $f(x) = y$. Since $x \in A \cup B$, we have either $x \in A$ or $x \in B$. If $x \in A$, then $y = f(x) \in f(A)$, whence $y \in f(A) \cup f(B)$; if $x \in B$, we deduce that $y = f(x) \in f(B)$, whence $y \in f(A) \cup f(B)$. Hence, $y \in f(A \cup B) \Rightarrow y \in f(A) \cup f(B)$, and hence

$$f(A \cup B) \subset f(A) \cup f(B).$$

We conclude that 5.3(i) holds.

Proof of 5.3(ii).—Since $A \cap B \subset A$ and $A \cap B \subset B$, we have (using 5.2)

$$f(A \cap B) \subset f(A) \cap f(B).$$

Proof of 5.3(iii).—Let $y \in f(A) - f(B)$. Then $y \in f(A)$ and $y \notin f(B)$. Hence, $y = f(x)$ with $x \in A$; since $y \notin f(B)$, it follows that $x \notin B$. Hence, $x \in A - B$, and hence $y = f(x) \in f(A - B)$. Therefore, $y \in f(A) - f(B) \Rightarrow y \in f(A - B)$. We conclude that 5.3(iii) holds.

We note here that it is not *necessarily* true that $f(A \cap B) = f(A) \cap f(B)$ for all f, $A \subset \operatorname{dom} f$ and $B \subset \operatorname{dom} f$. For instance, let $f : Z \to Z$ be defined by

$$f(x) = 0 \quad \text{for all} \quad x \in Z.$$

Let $A = \{0, 1\}$ and $B = \{2, 3, 4\}$; then

$$f(A \cap B) = f(\varnothing) = \varnothing \neq \{0\} = f(A) \cap f(B).$$

We leave it to the reader to show that it is not necessarily true that $f(A) - f(B) = f(A - B)$ for all f, $A \subset \operatorname{dom} f$ and $B \subset \operatorname{dom} f$.

Again, let $f = (X, Y, G)$ be a function. For every $B \subset Y$, we shall write

5.4 $$f^{-1}(B) = \{x \mid f(x) \in B\}.$$

Hence, $f^{-1}(B) \subset X$, and an object, x, belongs to $f^{-1}(B)$ if and only if $x \in X$ and $f(x) \in B$.

The set $f^{-1}(B)$ is called the inverse image of B by f. If $b \in Y$, then we shall write sometimes $f^{-1}(b)$ instead of $f^{-1}(\{b\})$.

Note that $f^{-1}(\varnothing) = \varnothing$.

Example 3.—Let f be the function in Example 2 of Chapter 4. Then
$$f^{-1}(\{a\}) = \{0, 1, 2\};$$
$$f^{-1}(\{c\}) = \{3\};$$
$$f^{-1}(\{b\}) = \varnothing.$$

Note that we may have $f^{-1}(B) = \varnothing$ even if $B \neq \varnothing$. Note also that if $y \in Y$, then $f^{-1}(\{y\})$ may contain more than one element; it may also happen that $f^{-1}(\{y\})$ is void.

It is immediate from the definition that if $f = (X, Y, G)$ is a function, and if $E \subset F \subset Y$, then

5.5 $$f^{-1}(E) \subset f^{-1}(F).$$

Among other properties that are useful and easy to establish, we mention the following.

5.6 *Let $f = (X, Y, G)$ be a function and let $A \subset Y$ and $B \subset Y$ be two sets. Then:*

(i) $f^{-1}(A \cup B) = f^{-1}(A) \cup f^{-1}(B)$;
(ii) $f^{-1}(A \cap B) = f^{-1}(A) \cap f^{-1}(B)$;
(iii) $f^{-1}(A - B) = f^{-1}(A) - f^{-1}(B)$;
(iv) $f^{-1}(\mathbf{C}_Y B) = \mathbf{C}_X f^{-1}(B)$.

Proof of 5.6(i).—Since $A \subset A \cup B$ and $B \subset A \cup B$, we have (using 5.5)

$$f^{-1}(A \cup B) \supset f^{-1}(A) \quad \text{and} \quad f^{-1}(A \cup B) \supset f^{-1}(B).$$

Hence
$$f^{-1}(A \cup B) \supset f^{-1}(A) \cup f^{-1}(B).$$

Now let $x \in f^{-1}(A \cup B)$. Then $f(x) \in A \cup B$. Hence, either $f(x) \in A$ and then $x \in f^{-1}(A)$, or $f(x) \in B$ and then $x \in f^{-1}(B)$. Therefore, in either case, $x \in f^{-1}(A) \cup f^{-1}(B)$. Hence

$$x \in f^{-1}(A \cup B) \Rightarrow x \in f^{-1}(A) \cup f^{-1}(B),$$

and hence

$$f^{-1}(A \cup B) \subset f^{-1}(A) \cup f^{-1}(B).$$

We conclude that 5.6(i) holds.

Proof of 5.6(ii).—Since $A \cap B \subset A$ and $A \cap B \subset B$, we have (using 5.5)

$$f^{-1}(A \cap B) \subset f^{-1}(A) \quad \text{and} \quad f^{-1}(A \cap B) \subset f^{-1}(B).$$

Whence

$$f^{-1}(A \cap B) \subset f^{-1}(A) \cap f^{-1}(B).$$

Now let $x \in f^{-1}(A) \cap f^{-1}(B)$. Then $x \in f^{-1}(A)$ and $x \in f^{-1}(B)$. Hence $f(x) \in A$ and $f(x) \in B$, and hence $f(x) \in A \cap B$. Hence $x \in f^{-1}(A \cap B)$. Therefore

$$x \in f^{-1}(A) \cap f^{-1}(B) \Rightarrow x \in f^{-1}(A \cap B),$$

whence

$$f^{-1}(A) \cap f^{-1}(B) \subset f^{-1}(A \cap B).$$

We conclude that 5.6(ii) holds.

Proof of 5.6(iii).—Let $x \in f^{-1}(A - B)$. Then $f(x) \in A - B$, whence $f(x) \in A$ and $f(x) \notin B$. Therefore

$$x \in f^{-1}(A) \quad \text{and} \quad x \notin f^{-1}(B),$$

hence

$$x \in f^{-1}(A) - f^{-1}(B).$$

Since $x \in f^{-1}(A - B)$ was arbitrary, we deduce

$$f^{-1}(A - B) \subset f^{-1}(A) - f^{-1}(B).$$

Now let $x \in f^{-1}(A) - f^{-1}(B)$. Then $x \in f^{-1}(A)$ and $x \notin f^{-1}(B)$, whence $f(x) \in A$ and $f(x) \notin B$. Therefore

$$f(x) \in A - B, \quad \text{hence} \quad x \in f^{-1}(A - B).$$

Since $x \in f^{-1}(A) - f^{-1}(B)$ was arbitrary, we deduce

$$f^{-1}(A) - f^{-1}(B) \subset f^{-1}(A - B).$$

We conclude that 5.6(iii) holds.

Proof of 5.6(iv).—This follows immediately from 5.6(iii).

Note that if $f = (X, Y, G)$ is a function and $A \subset Y$, $B \subset Y$, then we have

$$f^{-1}(A \cap B) = f^{-1}(A) \cap f^{-1}(B)$$

and

$$f^{-1}(A - B) = f^{-1}(A) - f^{-1}(B).$$

However, as we have seen, we *do not* necessarily have

$$f(A \cap B) = f(A) \cap f(B) \quad \text{and} \quad f(A - B) = f(A) - f(B)$$

if $A \subset X$, $B \subset X$ (see Exercise 5).

Exercise.—Let $f : X \to Y$ be a function, $A \subset X$ and $B \subset Y$. Then

$$f(f^{-1}(B)) \subset B \quad \text{and} \quad f^{-1}(f(A)) \supset A.$$

Show that "$\subset$ and $\supset$" cannot be replaced by "$=$" for all f, A and B. If $B \subset f(X)$, then $f(f^{-1}(B)) = B$ (see Exercise 6).

A function $f : X \to Y$ is said to be *surjective* (or *a surjection*) if

5.7 $f(X) = Y.$

Instead of saying that f is a surjection, we may say that f is a mapping of X *onto* Y.

A function $f : X \to Y$ is said to be *injective* (or *an injection*) if

5.8 $x' \in X, x'' \in X, x' \neq x'' \Rightarrow f(x') \neq f(x'').$

Instead of saying that f is an injection, we may say that f is a *one-to-one* mapping. Notice that a constant function cannot be injective if its domain contains more than one object.

A function $f : X \to Y$ is said to be *bijective* (or *a bijection*) if f is *both* a surjection and an injection. Hence,

5.9 f is a bijection $\Leftrightarrow f$ is an injection and a surjection.

Let X be a set. A bijection $\sigma : X \to X$ is sometimes called a *permutation* of X. Note that if σ is a permutation of X, then

$$\text{dom } \sigma = \text{rng } \sigma = X.$$

The set of all permutations of X will be denoted Σ_X; clearly $\Sigma_X \subset \mathcal{F}(X, X)$.

If $X = \{1, 2, \ldots, n\}$ $(n \in N)$ and if $\sigma \in \Sigma_X$ is defined by

$$\sigma(j) = a_j \quad (1 \leq j \leq n),$$

we shall write

5.10
$$\sigma = \curlyvee \begin{pmatrix} 1 & 2 & \dots & n \\ a_1 a_2 & \dots & a_n \end{pmatrix}.$$

This notation is often useful. If

$$\sigma = \curlyvee \begin{pmatrix} 1 & 2 & 3 \\ 3 & 2 & 1 \end{pmatrix},$$

then $\sigma \in \Sigma_{\{1,2,3\}}$ and $\sigma(1) = 3$, $\sigma(2) = 2$, $\sigma(3) = 1$.

Example 4.—Let X and Y be two sets. Consider the mapping $f \mapsto G_f$ of $\mathscr{F}(X, Y)$ into $\mathscr{P}(X \times Y)$. Clearly, this mapping is *injective*. In fact, if $f \in \mathscr{F}(X, Y)$ and $g \in \mathscr{F}(X, Y)$, then (see 4.2)

$$f = (X, Y, G') \quad \text{and} \quad g = (X, Y, G'').$$

Hence $f \neq g \Rightarrow G' \neq G'' \Rightarrow G_f \neq G_g$.

▼   The injection $f \mapsto G_f$, of $\mathscr{F}(X, Y)$ into $\mathscr{P}(X \times Y)$, identifies $\mathscr{F}(X, Y)$ with a *subset* of $\mathscr{P}(X \times Y)$.   ▲

The results of 5.3(i), 5.3(ii), 5.6(i), and 5.6(ii) can be easily generalized to n ($n \in N$) sets. Let $f : X \to Y$ be a function. Then, if $A_1, A_2, \dots, A_n$ are n subsets of X, we have:

5.7 $f(A_1 \cup A_2 \cup \dots \cup A_n) = f(A_1) \cup f(A_2) \cup \dots \cup f(A_n)$;

5.8 $f(A_1 \cap A_2 \cap \dots \cap A_n) \subset f(A_1) \cap f(A_2) \cap \dots \cap f(A_n)$.

If $B_1, B_2, \dots, B_n$ are n subsets of Y, we have:

5.9 $f^{-1}(B_1 \cup B_2 \cup \dots \cup B_n)$
$$= f^{-1}(B_1) \cup f^{-1}(B_2) \cup \dots \cup f^{-1}(B_n);$$

5.10 $f^{-1}(B_1 \cap B_2 \cap \dots \cap B_n)$
$$= f^{-1}(B_1) \cap f^{-1}(B_2) \cap \dots \cap f^{-1}(B_n).$$

We shall return to these generalizations later.

Exercises for Chapter 5

1. Let $X = \{0, 1, 2, 3, 4\}$, $Y = \{a, b, c, d\}$,
$$f = (X, Y, \{(0, a), (1, a), (2, b), (3, c), (4, c)\}).$$

Construct the following sets:

(a) $f(\{0, 3\})$;
(b) $f(\varnothing)$;
(c) $f^{-1}(b)$;
(d) $f^{-1}(\{a, c\})$;
(e) $f(\{1\} \cap \{2\})$;
(f) $f^{-1}(\varnothing - Y)$.

2. In 5.1, a function $f = (X, Y, G_f)$ is used to define another function with domain $\mathscr{P}(X)$. Write the triple that is this new function.

3.* Denote by F the triple (function) you wrote for Exercise 2. Denote by H the function

$$(\mathscr{P}(Y), \mathscr{P}(X), \{(B, f^{-1}(B)) \mid B \subset Y\}).$$

Show that:

(a) H is surjective $\Leftrightarrow f$ is injective $\Leftrightarrow F$ is injective;
(b) H is injective $\Leftrightarrow f$ is surjective $\Leftrightarrow F$ is surjective;
(c) H is bijective $\Leftrightarrow f$ is bijective $\Leftrightarrow F$ is bijective.

4. Let σ be the permutation of $\{1, 2, 3, 4, 5\}$ given by

$$\begin{pmatrix} 1 & 2 & 3 & 4 & 5 \\ 2 & 1 & 4 & 5 & 3 \end{pmatrix}.$$

(a) Write out the sets $\sigma(\{1, 3\})$, $\sigma^{-1}(\{2, 3, 4\})$ and $\sigma^{-1}(\{1\}) \cup \sigma(\{2\})$;
(b) Construct the set $\sigma(\{1\}) \cup \sigma(\{2\})$.

5.* Let $f = (X, Y, G)$ be a function. Show that:

(a) $f(A \cap B) = f(A) \cap f(B)$ for all $A \subset X$ and $B \subset X$ if and only if f is injective;
(b) $f(A - B) = f(A) - f(B)$ for all $A \subset X$ and $B \subset X$ if and only if f is injective;
(c) $f(\mathbf{C}A) = \mathbf{C}f(A)$ for all $A \subset X$ if and only if f is bijective.

6.* Let $f = (X, Y, G)$ be a function. Show that:

(a) $f(f^{-1}(B)) = B$ for all $B \subset Y$ if and only if f is surjective;
(b) $f^{-1}(f(A)) = A$ for every $A \subset X$ if and only if f is injective.

Chapter 6

Composition of Functions

Let $f: X \to Y$ and $g: Y \to Z$ be two functions. We shall define a new function $h: X \to Z$ by

$$h(x) = g(f(x)) \quad \text{for} \quad x \in X.$$

The function h will be denoted by $g \circ f$ and called the *composition* of g and f. The notation $g \circ f$ is read "g composition f."

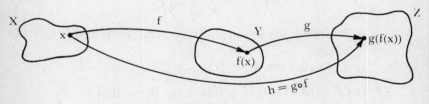

FIGURE 6

Example 1.—Let $f: R \to R$ and $g: R \to R$ be defined by

$$f(x) = 100 + x \quad \text{for} \quad x \in R$$

and

$$g(x) = x^{100} \quad \text{for} \quad x \in R.$$

Then

$$g \circ f(x) = (100 + x)^{100} \quad \text{for} \quad x \in R.$$

If $f: X \to Y$, $g: Y \to Z$, and $h: Z \to T$ are functions, then

6.1
$$h \circ (g \circ f) = (h \circ g) \circ f.$$

In fact, let $x \in X$. Then

$$h \circ (g \circ f)(x) = h(g \circ f(x)) = h(g(f(x)))$$

and

$$(h \circ g) \circ f(x) = h \circ g(f(x)) = h(g(f(x))).$$

Since $x \in X$ was arbitrary, we deduce 6.1 (see 4.4).

For f, g, and h, as above, we may write $h \circ g \circ f$ instead of $h \circ (g \circ f)$ or $(h \circ g) \circ f$.

6.2   *Now let $f : X \to Y$ and $g : Y \to Z$ be two functions and let $A \subset X$ and $C \subset Z$. Then*

(i) $g \circ f(A) = g(f(A))$;
(ii) $(g \circ f)^{-1}(C) = f^{-1}(g^{-1}(C))$.

Proof of 6.2(i).—Let $z \in g \circ f(A)$. Then there is $x \in A$ such that $z = g(f(x))$. Since $f(x) \in f(A)$, we have $z \in g(f(A))$. Since $z \in g \circ f(A)$ was arbitrary, we deduce

$$g \circ f(A) \subset g(f(A)).$$

Now let $z \in g(f(A))$. Then there is $y \in f(A)$ such that $z = g(y)$. Since $y \in f(A)$, there is $x \in A$ such that $y = f(x)$. Hence

$$z = g(y) = g(f(x)) = g \circ f(x) \Rightarrow z \in g \circ f(A).$$

Since $z \in g(f(A))$ was arbitrary, we deduce

$$g(f(A)) \subset g \circ f(A).$$

We conclude that 6.2(i) holds.

Proof of 6.2(ii).—Let $x \in (g \circ f)^{-1}(C)$. Then $g \circ f(x) \in C$; that is, $g(f(x)) \in C$. Hence, $f(x) \in g^{-1}(C)$, and hence $x \in f^{-1}(g^{-1}(C))$. Since $x \in (g \circ f)^{-1}(C)$ was arbitrary, we deduce

$$(g \circ f)^{-1}(C) \subset f^{-1}(g^{-1}(C)).$$

Let $x \in f^{-1}(g^{-1}(C))$. Then $f(x) \in g^{-1}(C)$ and $g(f(x)) \in C$. Since $g \circ f(x) = g(f(x)) \in C$, we have $x \in (g \circ f)^{-1}(C)$. Since $x \in f^{-1}(g^{-1}(C))$ was arbitrary, we deduce

$$f^{-1}(g^{-1}(C)) \subset (g \circ f)^{-1}(C).$$

We conclude that 6.2(ii) holds.

6.3   *Let $f : X \to Y$ and $g : Y \to Z$ be two functions. Then:*

(i) *$g \circ f$ is a surjection if g and f are surjections;*
(ii) *$g \circ f$ is an injection if g and f are injections;*
(iii) *$g \circ f$ is a bijection if g and f are bijections.*

Proof of 6.3(i).—By 6.2(i), we have

$$g \circ f(X) = g(f(X)) = g(Y) = Z,$$

whence $g \circ f$ is surjective.

Proof of 6.3(ii).—Let $x' \in X$, $x'' \in X$ such that $x' \neq x''$. Since f is injective, we have $f(x') \neq f(x'')$. Since g is also injective, $g(f(x')) \neq g(f(x''))$; that is, $g \circ f(x') \neq g \circ f(x'')$. Hence, $g \circ f$ is injective.

Proof of 6.3(iii).—The assertion in 6.3(iii) follows from 6.3(i) and 6.3(ii).

Remark.—If f and g are as in 6.3, and *if $g \circ f$ is an injection, then f is necessarily an injection.* In fact, suppose that $g \circ f$ is injective and that f is *not* injective. Then there are $x' \in X$, $x'' \in X$, $x' \neq x''$ such that $f(x') = f(x'')$. Hence $g(f(x')) = g(f(x''))$; that is, $g \circ f(x') = g \circ f(x'')$. Clearly, this is a contradiction, since $g \circ f$ is injective. Since the assumption that f is not injective leads to a contradiction, it follows that f is injective.

We now introduce some useful notations: Suppose, for instance, that $f : X \to Y_1$ and $g : X \to Y_2$ are two functions. The mapping

6.4 $x \mapsto (f(x), g(x))$

of X into $Y_1 \times Y_2$ will be denoted usually by (f, g).*

If $\varphi : Y_1 \times Y_2 \to Y$, then we shall sometimes write $\varphi(f, g)$ instead of $\varphi \circ (f, g)$. Hence

$$\varphi(f, g)(x) = \varphi(f(x), g(x))$$

for all $x \in X$. We can use similar notations for more than two functions. For instance, if f, g, h, k are the functions, on R to R, defined by

$$f(x) = x, \; g(x) = x^2, \; h(x) = x^3, \; k(x) = x^4$$

for $x \in R$, then (f, g, h, k) is the mapping

$$x \mapsto (x, x^2, x^3, x^4)$$

of R into $R^4 = R \times R \times R \times R$. If $\varphi : R^4 \to R$ is defined by

$$\varphi(x, y, z, t) = x + 2y + 3z + 4t$$

---

* We note that the same notation was used for representing the couple formed with the objects f and g. Clearly the couple(f, g) is not the same thing as the function (f, g). The meaning of the notation will always be clear from the context.

for $(x, y, z, t) \in \mathbf{R}^4$, then $\varphi(f, g, h, k)$ is the mapping

$$x \mapsto x + 2x^2 + 3x^3 + 4x^4$$

of $\mathbf{R}$ into $\mathbf{R}$.

If $f : X_1 \to Y_1$ and $g : X_2 \to Y_2$ are two functions, then the mapping

$$(x_1, x_2) \mapsto (f(x_1), g(x_2))$$

is usually denoted $f \times g$.

We have already defined (following the proof of 6.1) $h \circ g \circ f$. Suppose now that $f_j : X_j \to X_{j+1}$ $(j = 1, 2, \ldots, n, n > 1)$ are n mappings. If we have already defined $f_{n-1} \circ \ldots \circ f_1$, then we define $f_n \circ f_{n-1} \circ \ldots \circ f_1$ by

$$f_n \circ f_{n-1} \circ \ldots \circ f_1 = f_n \circ (f_{n-1} \circ \ldots \circ f_1).$$

The mapping $f_n \circ f_{n-1} \circ \ldots \circ f_1$ has domain X_1 and range X_{n+1}; it is called the composition of the mappings $f_n, f_{n-1}, \ldots, f_1$.

The following result, which we state without proof, is a generalization of 6.1.

If $f_1, f_2, \ldots, f_n$ are as above, and if $1 \leq p < n$, then*

6.5 $\quad f_n \circ \ldots \circ f_1 = (f_n \circ \ldots \circ f_{p+1}) \circ (f_p \circ \ldots \circ f_1).$

This assertion can be justified by mathematical induction (see Chap. 12).

Let $f : X \to Y$ be a mapping and let $A \subset X$. We may then consider the function $f_1 : A \to Y$ defined by

$$f_1(x) = f(x) \quad \text{for} \quad x \in A.$$

Clearly, $\operatorname{dom} f_1 = A$ and $\operatorname{rng} f_1 = Y$. The function f_1 is called the *restriction* of f to A; it is usually denoted $f \,|\, A$.

As we noted above, f and $f \,|\, A$ differ only in that they have distinct domains. Clearly, $f = f \,|\, X$.

For every set X and $A \subset X$ denote by $j_{A,X}$ the function on A to X defined by

$$j_{A,X}(x) = x \quad \text{for all} \quad x \in A.$$

Then

6.6 $\qquad\qquad\qquad f \,|\, A = f \circ j_{A,X}.$

---

* In 6.5, if $p + 1 = n$ we replace $f_n \circ \ldots \circ f_{p+1}$ by f_n; if $p = 1$ we replace $f_p \circ \ldots \circ f_1$, by f_1.

In fact, $f \mid A$ and $f \circ j_{A,x}$ both have domain A and range Y. If $x \in A$, then

$$(f \mid A)(x) = f(x) \quad \text{and} \quad f \circ j_{A,x}(x) = f(j_{A,x}(x)) = f(x),$$

hence (see 4.4), 6.6 is proved.

If $A = X = Y$, then, as we indicated in Example 6 of Chapter 4, the function $j_{A,A}$ is often denoted j_A.

We shall close this section with the following remarks: Let $f : X \rightarrow Y$ and $g : W \rightarrow Z$ be two functions. Note that we defined the composition $g \circ f$ only in the case in which $Y = W$. We may, however, weaken this condition. In fact, if

6.7 $$f(X) \subset W,$$

then we may still define $g \circ f$ by

$$g \circ f(x) = g(f(x)) \quad \text{for} \quad x \in X.$$

Note that 6.7 means that

6.8 $$x \in X \Rightarrow f(x) \in W.$$

Example 2.—Let $f : \boldsymbol{R} \rightarrow \boldsymbol{R}$ be defined by

$$f(x) = x^2 + 1$$

for $x \in \boldsymbol{R}$. Let $\boldsymbol{R}^* = \{x \mid x \in \boldsymbol{R}, x \neq 0\}$ and let $g : \boldsymbol{R}^* \rightarrow \boldsymbol{R}$ be defined by

$$g(x) = 1/x$$

for $x \in \boldsymbol{R}^*$. Clearly, 6.8 is satisfied, since $x^2 + 1 \neq 0$ for $x \in \boldsymbol{R}$. Hence, $g \circ f$ can be defined (as a mapping of $\boldsymbol{R}$ into $\boldsymbol{R}$):

$$g \circ f(x) = 1/(x^2 + 1) \quad \text{for} \quad x \in \boldsymbol{R}.$$

Exercises for Chapter 6

1. Let $X = Y = Z = \{1, 2, 3, 4\}$, and define:
$$f = (X, Y, \{(1, 1), (2, 1), (3, 1), (4, 2)\});$$
$$g = (Y, Z, \{(1, 4), (3, 2), (2, 3), (4, 3)\}).$$
Construct the functions $g \circ f$ and $f \circ g$.

2. If $f = (X, Y, G)$ and $f' = (X', Y', G')$ are functions such that $X' \supset X$, $Y' \supset Y$, and $G' \supset G$, we say that f' is an *extension* of f. Suppose $h = (A, B, H)$ and $h' = (A', B', H')$ are functions such that $A \subset A'$ and $B \subset B'$. Show that h' is an extension of $h \Leftrightarrow h' \circ j_{A,A'} = j_{B,B'} \circ h$.

3.* Let $f = (X, Y, G)$ be a function. Suppose that $X \cap \mathscr{P}(X) = \varnothing$. Show that

$$\varphi = (X \cup \mathscr{P}(X), Y \cup \mathscr{P}(Y), G \cup \{(A, f(A)) \mid A \subset X\})$$

is an extension of f.

4. Let $f = (\{0\}, \{0\}, \{(0, 0)\})$ and $g = (\{0\}, \{0, 1\}, \{(0, 0), (0, 1)\})$. Is g an extension of f?

5.* Let X be a set. Is

$$(\mathscr{F}(X, X) \times \mathscr{F}(X, X), \mathscr{F}(X, X), \{((f, g), f \circ g) \mid (f, g) \in \mathscr{F}(X, X)\})$$

a function?

6.* Let N denote the function $(\{0, 1\}, \{0, 1\}, \{(0, 1), (1, 0)\})$. Let S be a set and suppose that τ is any function with domain S and range $\{0, 1\}$. In Chapter 4, Example 11 and Exercise 9, we discussed the functions represented by the tables

M	0	1
0	0	1
1	1	1

m	0	1
0	0	0
1	0	1

C	0	1
0	1	1
1	0	1

Let σ be the function defined by the table

σ	0	1
0	1	1
1	1	0

Show that:

(a) $\quad N \circ \tau = \sigma \circ (\tau, \tau)$;

(b) $M \circ (\tau \times \tau) = \sigma \circ (\sigma \circ (\tau, \tau) \times \sigma \circ (\tau, \tau))$;

(c) $m \circ (\tau \times \tau) = \sigma \circ (\sigma \circ (\tau \times \tau), \sigma \circ (\tau \times \tau))$;

(d) $C \circ (\tau \times \tau) = M \circ ((N \circ \tau) \times \tau)$;

(e) $C \circ (\tau \times \tau) = M \circ m \circ (\tau \times (N \circ \tau))$;

(f) $C \circ (\tau \times \tau) = \sigma \circ (\sigma \circ (\tau, \tau) \times \sigma \circ (\tau, \tau)) \circ (\sigma \circ (\tau, \tau) \times \tau)$.

7.* Let X, Y, and Z be sets and suppose $f = (X, Y, F)$ and $g = (Y, Z, G)$ are functions. Show that $g \circ f = (X, Z, G \circ F)$ (see Exercise 5, Chap. 3).

Chapter 7

The Inverse Function

Recall that if A is a set, then j_A is the function $x \mapsto x$ on A to A. The following theorem is a useful one.

7.1 Theorem.—*Let $f : X \to Y$ be a mapping. Suppose that there are two mappings, $g : Y \to X$ and $h : Y \to X$, such that*

7.2 $$g \circ f = j_X \quad \text{and} \quad f \circ h = j_Y.$$

Then f is a bijection and $g = h$.

Note that

$$g \circ f = j_X \Leftrightarrow g(f(x)) = x \quad \text{for all} \quad x \in X$$

and

$$f \circ h = j_Y \Leftrightarrow f(h(y)) = y \quad \text{for all} \quad y \in Y.$$

Proof.—Let $x' \in X$, $x'' \in X$, $x' \neq x''$. Then

$$g(f(x')) = j_X(x') = x' \neq x'' = j_X(x'') = g(f(x'')).$$

Since $g(f(x')) \neq g(f(x''))$, we deduce $f(x') \neq f(x'')$; hence f is an *injection*.

Now let $y \in Y$; then $h(y) \in X$ and

$$f(h(y)) = f \circ h(y) = j_Y(y) = y.$$

Hence, given $y \in Y$, the element $x = h(y) \in X$ satisfies $f(x) = y$; whence f is a *surjection*. Hence, f is both an injection and a surjection; therefore it is a bijection.

Note now that

$$g(y) = g(f \circ h(y)) = (g \circ f)(h(y)) = j_X(h(y)) = h(y)$$

for all $y \in Y$; we conclude $g = h$.

Hence, 7.1 is completely proved.

Remark.—Obviously, Theorem 7.1 shows also that g ($g = h$) is a bijection (why?).

7.3 Theorem.—*Let $f : X \to Y$ be a bijection. Then there is a unique function $g : Y \to X$ such that:*

7.4 $$g(f(x)) = x \quad \text{for} \quad x \in X;$$

7.5 $$f(g(y)) = y \quad \text{for} \quad y \in Y.$$

The function g is a bijection.

Proof.—Once the existence of g is proved, then the uniqueness assertion and the fact that g is a bijection follow from Theorem 7.1 (see the remark following Theorem 7.1).

Hence, to prove Theorem 7.3, it remains to prove the existence of the function g. For this, we reason as follows: Let $y \in Y$. Since $f(X) = Y$, there is $x \in X$ satisfying $f(x) = y$; since f is an injection, there is only one such $x \in X$. Denote this unique $x \in X$ by $g(y)$. We define in this way a mapping $y \mapsto g(y)$ of Y into X. From its definition, it follows that

$$f(g(y)) = y \quad \text{for all} \quad y \in Y.$$

Furthermore if $x \in X$ and $y = f(x)$, then x is the unique element in X satisfying $y = f(x)$; whence $x = g(y)$; that is

$$g(f(x)) = g(y) = x.$$

Hence, Theorem 7.3 is proved.

Let now $f : X \to Y$ be a bijection. *The unique function $g : Y \to X$ satisfying*

7.6 $$g(f(x)) = x \quad \text{for} \quad x \in X$$

and

7.7 $$f(g(y)) = y \quad \text{for} \quad y \in Y$$

is called the function inverse to f (or the inverse function of f).

The function inverse to f is denoted f^{-1}; with this notation, 7.6 and 7.7 become, respectively

7.8 $$f^{-1}(f(x)) = x \quad \text{for} \quad x \in X$$

and

7.9 $f(f^{-1}(y)) = y$ *for* $y \in Y$.

Note that

$$X \overset{f}{\to} Y \quad \text{and} \quad X \overset{f^{-1}}{\leftarrow} Y.$$

Note also that f^{-1} was defined above *only* for $f: X \to Y$ a bijection, and that $f^{-1}: Y \to X$ is also a bijection.

Therefore, if $f: X \to Y$ is a bijection, and $g: Y \to X$ satisfies 7.6 and 7.7, then $g = f^{-1}$. This remark will be used often in what follows.

Example 1.—Let $f: R \to R$ and $g: R \to R$ be defined by $f(x) = x + 1$ for $x \in R$ and $g(x) = x - 1$ for $x \in R$. Then

$$g(f(x)) = g(x + 1) = (x + 1) - 1 = x \quad \text{for} \quad x \in R$$

and

$$f(g(y)) = f(y - 1) = (y - 1) + 1 = y \quad \text{for} \quad y \in R.$$

We conclude that $g: R \to R$ is a bijection and that $g = f^{-1}$.

Example 2.—Let $R^* = \{x \mid x \in R, x \neq 0\}$ and let $f: R^* \to R^*$ and $g: R^* \to R^*$ be defined by $f(x) = g(x) = 1/x$ for $x \in R^*$. Then

$$g(f(x)) = g(1/x) = 1/(1/x) = x \quad \text{for} \quad x \in R^*.$$

Since $g = f$, we also have $f(g(x)) = x$ for $x \in R^*$. We conclude that $f: R^* \to R^*$ is a bijection and that $f^{-1} = f$.

Example 3.—Let A and B be two sets and let f be the function $(a, b) \mapsto (b, a)$, on $A \times B$ to $B \times A$. Then f is a bijection and f^{-1} is the function $(b, a) \mapsto (a, b)$, on $B \times A$ to $A \times B$.

Remarks.—(1) If $f: X \to Y$ is an injection, and if there is a mapping $g: Y \to X$ such that 7.7 is satisfied (that is, $f(g(y)) = y$ for $y \in Y$), then we deduce that f is a surjection (why?). Hence, f is a bijection, and then 7.1 shows that $g = f^{-1}$.

(2) Let $f: X \to Y$ be an injection. We noted in (1) that if we can show that there is $g: Y \to X$ such that $f(g(y)) = y$ for $y \in Y$, then f is a bijection and $g = f^{-1}$. Hence, if we can find "a solution

$g: Y \to X$ of the equation $f \circ g = j_Y$," we conclude that f is a bijection.

For instance, let f be the mapping $x \mapsto 2x + 3$ of R into R. Clearly, f is an injection. A mapping $g: R \to R$ satisfies $f(g(x)) = x$ for $x \in R$ if and only if

$$2g(x) + 3 = x \quad \text{for} \quad x \in R.$$

Now, for $x \in R$,

$$2g(x) + 3 = x \Leftrightarrow g(x) = \tfrac{1}{2}(x - 3).$$

Since the mapping $g: x \mapsto \tfrac{1}{2}(x - 3)$ of R into R satisfies $f \circ g = j_Y$, we deduce that f is a bijection and $f^{-1} = g$.

7.10 Theorem.—*Let $f: X \to Y$ and $g: Y \to Z$ be two bijections. Then $g \circ f$ is a bijection and*

$$(g \circ f)^{-1} = f^{-1} \circ g^{-1}.$$

Proof.—By 6.3, $g \circ f$ is a bijection. Also, $u = f^{-1} \circ g^{-1}$ is a well-defined mapping of Z into X. Furthermore

$$u \circ (g \circ f) = (f^{-1} \circ g^{-1}) \circ (g \circ f) = f^{-1} \circ (g^{-1} \circ g) \circ f$$
$$= f^{-1} \circ j_Y \circ f = f^{-1} \circ f = j_X$$

and

$$(g \circ f) \circ u = (g \circ f) \circ (f^{-1} \circ g^{-1}) = g \circ (f \circ f^{-1}) \circ g^{-1}$$
$$= g \circ j_X \circ g^{-1} = g \circ g^{-1} = j_Y.$$

Hence, 7.10 is proved.

We close this section with the following remarks: Let X and Y be two sets, let $A \subset X$, and let $f: A \to Y$ be a function. Suppose that there is a function $g: f(A) \to X$ such that:

7.11 $\qquad\qquad g(f(x)) = x \quad \text{for} \quad x \in A;$

7.12 $\qquad\qquad f(g(x)) = x \quad \text{for} \quad x \in f(A).$

(Note that 7.12 implies that $g(y) \in A$ for $y \in f(A)$; in fact, $\mathrm{dom}\, f = A$, and hence $f(z)$ is defined only for $z \in A$.) In this case, we shall also say that g is the *function inverse* to f or the *inverse function* of f. Note that g considered as a function on $f(A)$ to A is the function inverse to the mapping $x \mapsto f(x)$ on A to $f(A)$.

Exercises for Chapter 7

1. Let $X = \{1, 2, 3\}$ and let σ be the bijection

$$(X, X, \{(1, 2), (2, 1), (3, 3)\}).$$

Express σ^{-1} as a triple.

2. Let $f = (X, Y, G)$ be a bijection. Express f^{-1} as a triple.

3.* Let $f = (X, X, G)$ be a function and suppose there is $n \in N$ such that $f^n = j_X$ (for $n \in N$ we denote $f^n = f$ if $n = 1$ and $f^n = f \circ \ldots \circ f$ (n times) if $n > 1$). Show that f is a bijection.

Chapter 8

Unions and Intersections of Families of Sets

A function $f = (I, X, G)$ is sometimes called a *family of elements of X.* In this case, we say that I is the set of *indices* of the family. We write f_i instead of $f(i)$ (for $i \in I$) and $f = (f_i)_{i \in I}$. The set

$$f(I) = \{f_i \mid i \in I\}$$

is called the set of elements of the family.

Notations like $(x_i)_{i \in I}$ are often used to represent a family of elements of a set X.*

Let $(x_i)_{i \in I}$ be a family of elements of X; for each $i \in I$, we call x_i *the term of index i* or the "*i*th component" of the family. If $(x_i)_{i \in I}$ is a family, and $J \subset I$, we often say that $(x_i)_{i \in J}$ is a *subfamily* of $(x_i)_{i \in I}$.

A *sequence* of elements of X is a family of elements of X having for set of indices *a part of* **Z**.

Notations such as those indicated below are often used for representing sequences:

$$(x_i)_{i \geq 1} \quad \text{instead of} \quad (x_i)_{i \in N};$$

$$(x_i)_{1 \leq i \leq p} \quad \text{instead of} \quad (x_i)_{i \in \{1, \ldots, p\}}.$$

Example 1.—(i) $(2n + 1)_{n \in N}$ is a sequence of elements of **R**; the term of index $n \in N$ of this sequence is the number $2n + 1$.

(ii) $(2n)_{n \in Z}$ is a sequence; the term of index 3 of this sequence is 6.

---

* When a family is represented by such a notation, $i \mapsto x_i$ is a mapping of I into X.

If Y is a set and $(A_i)_{i \in I}$ is a family of elements of $\mathscr{P}(Y)$, we often say that $(A_i)_{i \in I}$ *is a family of subsets (or parts) of Y* (or simply a family of sets).

In Chapter 2 we defined the union and the intersection of two sets. We shall now define the union and the intersection of a family of sets.

Union.—Let X be a set and let $(A_i)_{i \in I}$ be a family of parts of X. We define the set $\bigcup\limits_{i \in I} A_i$ by

8.1 $x \in \bigcup\limits_{i \in I} A_i \Leftrightarrow$ there exists $i \in I$ such that $x \in A_i$.

Hence, an object, $x \in X$, belongs to $\bigcup\limits_{i \in I} A_i$ if and only if x belongs to at least one A_i, $i \in I$. If $I = \varnothing$,

8.2
$$\bigcup\limits_{i \in I} A_i = \varnothing.$$

The set $\bigcup\limits_{i \in I} A_i$ is called the union of the family $(A_i)_{i \in I}$; it is read "union i in I, A_i."

Intersection.—Let X be a set and let $(A_i)_{i \in I}$ be a family of parts of X. We define the set $\bigcap\limits_{i \in I} A_i$ by

8.3 $x \in \bigcap\limits_{i \in I} A_i \Leftrightarrow x \in X$ and $x \in A_i$ for *all* $i \in I$.

Hence, an object $x \in X$ belongs to $\bigcap\limits_{i \in I} A_i$ if and only if the condition $x \in A_i$ is satisfied for all $i \in I$. If $I = \varnothing$,

8.4
$$\bigcap\limits_{i \in I} A_i = X.$$

The set $\bigcap\limits_{i \in I} A_i$ is called the intersection of the family $(A_i)_{i \in I}$; it is read "intersection i in I, A_i."

Example 2.—(i) For each $p \in N$, let $A_p = \{n \mid n \in N, n \geq p\}$; then $\bigcap\limits_{p \in N} A_p = \varnothing$.

(ii) Let X be a set containing at least two elements; let $a \in X$ and $I = X - \{a\}$. For each $x \in I$, let $A_x = X - \{x\}$. Then $\bigcap\limits_{x \in I} A_x = \{a\}$.

▼   The set $\bigcup\limits_{i \in I} A_i$ does not depend on X. The set $\bigcap\limits_{i \in I} A_i$ does not depend on X either, if $I \neq \varnothing$. It does depend on X (see 8.4) if $I = \varnothing$.   ▲

Notations such as those indicated below are often used in the case of sequences:

$$\bigcup_{i \geq 1} A_i \quad \text{instead of} \quad \bigcup_{i \in N} A_i;$$

$$\bigcup_{1 \leq i \leq p} A_i \quad \text{instead of} \quad \bigcup_{i \in \{1, \dots, p\}} A_i.$$

It is easy to see that the definitions given in Chapter 2 are particular cases of 8.1 and 8.3.

For instance, let A and B be two subsets of X and let $I = \{A, B\}$. Let $X_A = A$ and $X_B = B$. Then, clearly

$$\bigcup_{S \in I} X_S = A \cup B \quad \text{and} \quad \bigcap_{S \in I} X_S = A \cap B.$$

Now let $A_1, \dots, A_n$ be n sets. The set $A_1 \cup \dots \cup A_n$, as defined in Chapter 2, coincides with $\bigcup_{i \in \{1, \dots, n\}} A_i$ as defined in 8.1; also, $A_1 \cap \dots \cap A_n$ as defined in Chapter 2 coincides with

$$\bigcap_{i \in \{1, \dots, n\}} A_i$$

as defined in 8.3. Hence, we may write

$$\bigcup_{i \in \{1, \dots, n\}} A_i = A_1 \cup \dots \cup A_n \quad \text{and} \quad \bigcap_{i \in \{1, \dots, n\}} A_i = A_1 \cap \dots \cap A_n.$$

Now let X be a set and $\mathscr{F} \subset \mathscr{P}(X)$ (hence $\mathscr{F}$ is a set of subsets of X). For each $A \in \mathscr{F}$, let $X_A = A$. Then $(X_A)_{A \in \mathscr{F}}$ is a *family*, the set of indices of which is $\mathscr{F}$. We write

8.5
$$\bigcup_{A \in \mathscr{F}} A = \bigcup_{A \in \mathscr{F}} X_A$$

and we call $\bigcup_{A \in \mathscr{F}} A$ the *union* of $\mathscr{F}$; we also write

8.6
$$\bigcap_{A \in \mathscr{F}} A = \bigcap_{A \in \mathscr{F}} X_A$$

and call $\bigcap_{A \in \mathscr{F}} A$ the *intersection* of $\mathscr{F}$.

▼   In this section, we defined the union and the intersection of a family of sets $(A_i)_{i \in I}$. Note that here we have always supposed that the sets $A_i (i \in I)$ are subsets of a *given* set. When we defined the union and the intersection of two sets in Chapter 2, we did not suppose *a priori* that the considered sets were subsets of a given set. It follows, however, from the usual axioms of set theory, that if

$\mathscr{F}$ *is a set* the elements of which are sets, then there exists a set X such that $\mathscr{F} \subset \mathscr{P}(X)$. It also follows from these axioms that if x and y are two "objects", then there exists a set the "objects" of which are just x and y.

The union, $\bigcup\limits_{i \in I} A_i$, can be defined without supposing *a priori* that A_i, $i \in I$, are subsets of a given set. The same is true for $\bigcap\limits_{i \in I} A_i$ if $I \neq \varnothing$. However, $\bigcap\limits_{i \in I} A_i$ is not defined if $I = \varnothing$. This is due to the fact that there is no *set* Y such that $y \in Y$ for *every* y.   ▲

We shall now discuss the Cartesian product of a family of sets. Let X be a set and let $(A_i)_{i \in I}$ $(I \neq \varnothing)$ be a family of parts of X. We define the set $\prod\limits_{i \in I} A_i$ by

8.7 $x \in \prod\limits_{i \in I} A_i \Leftrightarrow x = (x_i)_{i \in I}$ with $x_i \in A_i$ for *all* $i \in I$.

Hence, $\prod\limits_{i \in I} A_i$ is the set of all families $(x_i)_{i \in I}$ of elements of X such that $x_i \in A_i$ for all $i \in I$. Hence, $\prod\limits_{i \in I} A_i$ is the set of all functions $f : I \to X$ such that

$$f(i) \in A_i \quad \text{for all} \quad i \in I.$$

It follows that if $A_i = X$ for all $i \in I$, then (see 4.3)

$$\prod\limits_{i \in I} A_i = X^I.$$

The set $\prod\limits_{i \in I} A_i$ is called the *Cartesian product* or the product of the family $(A_i)_{i \in I}$; it is read "product $i \in I$, A_i." If $I = \{1\}$, $I = \{1, 2\}$, or $I = \{1, 2, \ldots, n\}$, then we identify $\prod\limits_{i \in I} A_i$ with A_1, $A_1 \times A_2$, or $A_1 \times A_2 \times \ldots \times A_n$ respectively.

For each $k \in I$, we denote by pr_k the mapping $(x_i)_{i \in I} \mapsto x_k$ of $\prod\limits_{i \in I} A_i$ into A_k; this mapping is called the *projection of index k*, or *the projection onto A_k*.

▼   The assertion, $\prod\limits_{i \in I} A_i \neq \varnothing$ if $A_i \neq \varnothing$ for *all* $i \in I$, is an axiom called the Axiom of Choice. It says, in fact, that if $(A_i)_{i \in I}$ is a family of non-void sets, then there *exists* a family $(x_i)_{i \in I}$ such that $x_i \in A_i$ for *all* $i \in I$.   ▲

We shall close this paragraph by listing (without proofs) several results concerning unions and intersections of families; they can easily be proved.

We denote, by X and Y, two sets.

8.8   Let $A \subset X$ and $(A_i)_{i \in I}$ a family of parts of X. Then:

(i) $A \cap (\bigcup_{i \in I} A_i) = \bigcup_{i \in I} (A \cap A_i)$;

(ii) $A \cup (\bigcap_{i \in I} A_i) = \bigcap_{i \in I} (A \cup A_i)$.

The results in 8.8 generalize assertions 2.7 and 2.8 (Chap. 2).

8.9   Let $(A_i)_{i \in I}$ be a family of parts of X. Then:

(i) $\mathbf{C}(\bigcup_{i \in I} A_i) = \bigcap_{i \in I} \mathbf{C}A_i$;

(ii) $\mathbf{C}(\bigcap_{i \in I} A_i) = \bigcup_{i \in I} \mathbf{C}A_i$.

The results in 8.9 generalize assertion 2.12.

8.10   Let $f : X \to Y$ and let $(A_i)_{i \in I}$ be a family of parts of X. Then:

(i) $f(\bigcup_{i \in I} A_i) = \bigcup_{i \in I} f(A_i)$;

(ii) $f(\bigcap_{i \in I} A_i) \subset \bigcap_{i \in I} f(A_i)$.

The results in 8.10 generalize assertion 5.3.

8.11   Let $f : X \to Y$ and let $(B_i)_{i \in I}$ be a family of parts of Y. Then:

(i) $f^{-1}(\bigcup_{i \in I} B_i) = \bigcup_{i \in I} f^{-1}(B_i)$;

(ii) $f^{-1}(\bigcap_{i \in I} B_i) = \bigcap_{i \in I} f^{-1}(B_i)$.

The results of 8.11 generalize 5.6.

We wish to point out some notations that are frequently encountered (note that $\{A \mid A \in \mathscr{F}\} = \mathscr{F}$):

$$\bigcup \{A \mid A \in \mathscr{F}\} \quad \text{instead of} \quad \bigcup_{A \in \mathscr{F}} A;$$

$$\bigcup \{A_i \mid i \in I\} \quad \text{instead of} \quad \bigcup_{i \in I} A_i;$$

$$\bigcap \{A \mid A \in \mathscr{F}\} \quad \text{instead of} \quad \bigcap_{A \in \mathscr{F}} A;$$

$$\bigcap \{A_i \mid i \in I\} \quad \text{instead of} \quad \bigcap_{i \in I} A_i.$$

▼   *Example 3.*—Let Z and T be two sets and let

$$X = \bigcup \{\mathscr{F}(A, T) \mid A \in \mathscr{P}(Z)\}.$$

An object of X is a function (A, T, G) with $A \in \mathscr{P}(Z)$. Let h be the mapping $(A, T, G) \mapsto A$ of X into $\mathscr{P}(Z)$ (thus $h(A, T, G) = pr_1(A, T, G)$). Clearly, then, for every $f \in X$, we have $h(f) = \operatorname{dom} f$. ▲

Exercises for Chapter 8

1. Let $A_1 = \{2, 1\}$, $A_2 = \{3, 2\}$, $A_3 = \{4, 3\}$ and in general $A_n = \{n + 1, n\}$. Write explicitly the sets:

(a) $\bigcup_{1 \leq i \leq 5} A_i$;

(b) $\bigcup_{i \in N} A_i$;

(c) $\bigcap_{i \in \{1,2\}} A_i$;

(d) $\bigcup_{i \geq 5} A_i$;

(e) $\bigcap_{i \in N} \mathbf{C} A_i$ (here $\mathbf{C} A_i = \mathbf{C}_N A_i$);

(f) $\bigcup_{i \in N} \mathbf{C} A_i$ (here $\mathbf{C} A_i = \mathbf{C}_N A_i$);

(g) $\prod_{i \in \{1,2\}} A_i$;

(h) $pr_8(\prod_{i \in N} A_i)$.

2. Write a bijection with domain $\prod_{i \in \{1,2\}} A_i$ and range $A_1 \times A_2$, where A_1 and A_2 are as above.

3. Let I and A be sets, and suppose $A_i \subset A$ for each $i \in I$. Show that

$$\{I\} \times \{A\} \times \mathscr{P}(I \times A) \supset A^I = \mathscr{F}(I, A) \supset \prod_{i \in I} A_i$$

4. Give a simple example of a set I, a set A, and a family $((A_i)_{i \in I})$ of subsets of A such that $\prod_{i \in I} A_i$ is properly contained in $\mathscr{F}(I, A)$; i.e., $\prod_{i \in I} A_i$ is not equal to $\mathscr{F}(I, A)$. Also, give an example in which $\prod_{i \in I} A_i = \mathscr{F}(I, A)$.

5. Let $f : N \to N$ be defined by $f(n) = n^2$ for each $n \in N$, i.e., $f = (N, N, \{(n, n^2) \mid n \in N\})$. Let A_n be defined as in Example 1

for $n \geq 1$. Write explicitly the sets:

(a) $\displaystyle\bigcup_{1 \leq i \leq 5} f(A_i)$;

(b) $\displaystyle\bigcap_{i \in N} B_i$ where $B_i = A_i \cup f(A_i)$, $i \in N$;

(c) $\displaystyle\bigcup_{i \in N} (A_i \cap f(A_i))$;

(d) $\displaystyle\bigcup_{i \geq 1} (A_i \cup \mathbf{C}f(A_i))$;

(e) $\displaystyle\bigcap_{i \geq 5} (A_i \cap \mathbf{C}f(A_i))$;

(f) $\displaystyle pr_k \left(\prod_{i \in N} f(A_i) \right)$, $k \in N$.

6. Let X be a set and let $\mathscr{G} \subset \mathscr{P}(X)$. Recall that $\bigcup \mathscr{G} = \bigcup_{G \in \mathscr{G}} G$ and $\bigcap \mathscr{G} = \bigcap_{G \in \mathscr{G}} G$. Let $X = \{1, 2, 3, 4\}$ and

$$\mathscr{G} = \{\{1, 2, 3\}, \{2, 3\}, \{1, 3, 4\}\};$$

determine $\bigcup \mathscr{G}$ and $\bigcap \mathscr{G}$. Let

$$X = N, \quad \mathscr{G} = \{\{n \mid n \geq 2k\} \mid k \in N\};$$

determine $\bigcup \mathscr{G}$ and $\bigcap \mathscr{G}$.

7.* Let $(J_\lambda)_{\lambda \in I}$ be a family of sets and let $J = \bigcup_{\lambda \in I} J_\lambda$. Let $(A_i)_{i \in J}$ be a family of parts of a set X. Then:

(a) $\displaystyle\bigcup_{i \in J} A_i = \bigcup_{\lambda \in I} \left(\bigcup_{i \in J_\lambda} A_i \right)$ (associativity);

(b) $\displaystyle\bigcap_{i \in J} A_i = \bigcap_{\lambda \in I} \left(\bigcup_{i \in J_\lambda} A_i \right)$ (associativity).

8.* Let $(A_i)_{i \in I}$ and $(B_i)_{i \in J}$ be two families of parts of X. Then:

(a) $\displaystyle\left(\bigcup_{i \in I} A_i \right) \cap \left(\bigcup_{j \in J} B_j \right) = \bigcup_{(i,j) \in I \times J} A_i \cap B_j$ (distributivity);

(b) $\displaystyle\left(\bigcap_{i \in I} A_i \right) \cup \left(\bigcap_{j \in J} B_j \right) = \bigcap_{(i,j) \in I \times J} A_i \cup B_j$ (distributivity).

Chapter 9

Equipotent Sets

Let A and B be two sets. We say that A and B are *equipotent* if there exists a bijection $f : A \to B$.

Note that if $f : A \to B$ is a bijection, then $f^{-1} : B \to A$ is a bijection too.

If A and B are equipotent and B and C are equipotent, then A and C are equipotent. In fact, since A and B are equipotent, there exists a bijection $f : A \to B$; since B and C are equipotent, there exists a bijection $g : B \to C$. Since $g \circ f : A \to C$ is a bijection (see 6.3(iii)), it follows that A and C are equipotent.

Example 1.—Let $A = \{5, 9, 10\}$ and $B = \{1, 2, 3\}$. Let $f : A \to B$ be defined by $f(5) = 1$, $f(9) = 2$, $f(10) = 3$. Then f is a bijection; hence, A and B are equipotent.

Example 2.—(i) Let $N' = \{2n \mid n \in N\}$ and let $f : N \to N'$ be the mapping defined by $f(n) = 2n$ for $n \in N$. Then f is a bijection; hence, N and N' are equipotent. (Note that although N and N' are equipotent, we have $N' \subset N$ and $N' \neq N$.)

(ii) Let $B = \{n \mid n \in N, n \geq 2\}$ and let $h : N \to B$ be the mapping defined by $h(n) = n + 1$ for $n \in N$. Then h is a bijection; hence, B and N are equipotent.

Example 3.—Let $h : Z \to N$ be defined by

$$h(n) = 2n + 1 \quad \text{if} \quad n = 0, 1, 2, \ldots \ ;$$
$$= -2n \quad \text{if} \quad n = -1, -2, \ldots \ .$$

Then $h : Z \to N$ is a bijection, whence N and Z are equipotent.

Example 4.—If A and B are sets and $f : A \to B$ is an injection then A and $f(A)$ are equipotent. In fact, the mapping $g : A \to f(A)$, defined by $g(x) = f(x)$ for $x \in A$, is a bijection.

▼ *Example 5.*—Let X and Y be two equipotent sets. Then Σ_X and Σ_Y are equipotent.

Let $f : X \to Y$ be a bijection and let $g = f^{-1}$. Now let $\varphi : \Sigma_Y \to \Sigma_X$ be defined by

$$\varphi(u) = g \circ u \circ f \ (X \xrightarrow{f} Y \xrightarrow{u} Y \xrightarrow{g} X)$$

for $u \in \Sigma_Y$. Since f, u, and g are bijections, it follows that $\varphi(u) \in \Sigma_X$ (see 6.3(iii)). Hence, $u \mapsto \varphi(u)$ *is a mapping of* Σ_Y *into* Σ_X.

If $v \in \Sigma_X$, then

$$u = f \circ v \circ g \ (Y \xrightarrow{g} X \xrightarrow{v} X \xrightarrow{f} Y)$$

belongs to Σ_Y, and

$$\varphi(u) = g \circ u \circ f = g \circ (f \circ v \circ g) \circ f$$
$$= (g \circ f) \circ v \circ (g \circ f) = v.$$

Hence, φ *is a surjection.*

Now let $u_1 \in \Sigma_Y$, $u_2 \in \Sigma_Y$, $u_1 \neq u_2$. Then there is $y \in Y$ such that $u_1(y) \neq u_2(y)$. Let $x \in X$, satisfy $f(x) = y$; then

$$u_1 \circ f(x) = u_1(y) \neq u_2(y) = u_2 \circ f(x);$$

therefore

$$g \circ u_1 \circ f(x) = g \circ u_1(y) \neq g \circ u_2(y) = g \circ u_2 \circ f(x);$$

that is, $g \circ u_1 \circ f \neq g \circ u_2 \circ f$. Thus, $\varphi(u_1) \neq \varphi(u_2)$; hence φ *is an injection.*

We conclude that φ is a bijection, and therefore that Σ_X and Σ_Y are equipotent. ▲

Example 6.—Let A, B, A', and B' be four sets. Suppose that A is equipotent to A' and B is equipotent to B'. Then $A \times B$ is equipotent to $A' \times B'$.

Since A is equipotent to A', there is a bijection $f : A \to A'$. Since B is equipotent to B', there is a bijection $g : B \to B'$. Let $h : A \times B \to A' \times B'$ be defined by

$$h(x, y) = (f(x), g(y))$$

for $(x, y) \in A \times B$. Then h is a bijection (why?). Hence, $A \times B$ and $A' \times B'$ are equipotent.

From the results in Examples 3 and 6, it follows that $Z \times Z$, $Z \times N$, $N \times Z$ and $N \times N$ are equipotent.

We shall say that a set S *is finite* if $S = \varnothing$ or if there exists $p \in N$ such that S is equipotent to $\{1, 2, \ldots, p\}$. Clearly, if S is finite and if S is equipotent to T, then T is finite also.

If $S = \varnothing$, we say that the number of elements of S is 0; if S is equipotent to $\{1, \ldots, p\}$ ($p \in N$), we say that* the number of elements of S is p. The number of elements of a finite set S is denoted $c(S)$.

If S is a finite set and $A \subset S$, then A and $S - A$ are finite and

9.1 $c(S) = c(A) + c(S - A)$.

A set S is said to be *countable* if it is equipotent to Z. If S is countable and if S is equipotent to T, then T is also countable. The result of Example 3 shows that N is countable.

A set S is said to be *infinite* if it is not finite. Clearly, if S is infinite and if S is equipotent to T, then T is also infinite. In what follows, we shall accept the following results:

9.2   *The set Z is infinite.*

9.3   *A set is infinite if and only if it contains a countable part.*

Theorem 9.4 is useful for showing that two sets are equipotent.

9.4 Bernstein-Schroeder Theorem.—*Let X and Y be two sets. Suppose that:*
   (i) *there exists an injection $f : X \to Y$;*
   (ii) *there exists an injection $g : Y \to X$.*
Then X and Y are equipotent.

▼   The proof of 9.4 is based on the following result, which will be established first:

9.5   *Let X be a set and let $\varphi : \mathscr{P}(X) \to \mathscr{P}(X)$ be a mapping satisfying the condition*
$$A \subset B \Rightarrow \varphi(A) \subset \varphi(B).$$
Then there is $D \in \mathscr{P}(X)$ such that $\varphi(D) = D$.

---

* Here (and in what follows) we shall accept certain results (which are intuitively obvious) concerning finite sets. For instance, we accept that: If $n \in N$, $m \in N$ and if $\{1, \ldots, n\}$ and $\{1, \ldots, m\}$ are equipotent, then $n = m$.

Proof.—Let $\mathcal{K} = \{S \mid S \subset X, S \subset \varphi(S)\}$ and define

9.6
$$D = \bigcup_{S \in \mathcal{K}} S.$$

Since $S \subset D$ for each $S \in \mathcal{K}$ we have, using the hypothesis on φ, $\varphi(S) \subset \varphi(D)$ for each $S \in \mathcal{K}$. From the definition of $\mathcal{K}$, we have $S \subset \varphi(S)$ for each $S \in \mathcal{K}$; thus $S \subset \varphi(D)$ for each $S \in \mathcal{K}$. Hence

$$\bigcup_{S \in \mathcal{K}} S \subset \varphi(D); \quad \text{that is,}$$

9.7
$$D \subset \varphi(D).$$

Using 9.7 and the hypothesis on φ again, we obtain $\varphi(D) \subset \varphi(\varphi(D))$, hence $\varphi(D) \in \mathcal{K}$. Therefore

9.8
$$\varphi(D) \subset D.$$

Comparing 9.7 and 9.8, we conclude $\varphi(D) = D$.

Proof of Theorem 9.4.—Let $\varphi: \mathscr{P}(X) \to \mathscr{P}(X)$ be the mapping defined by $\varphi(S) = g(\mathbf{C}f(\mathbf{C}S))$ for $S \in \mathscr{P}(X)$. Then

$$A \subset B \Rightarrow \mathbf{C}A \supset \mathbf{C}B \Rightarrow f(\mathbf{C}A) \supset f(\mathbf{C}B) \Rightarrow \mathbf{C}f(\mathbf{C}A)) \subset \mathbf{C}f(\mathbf{C}B)).$$

Hence,

$$\varphi(A) = g(\mathbf{C}f(\mathbf{C}A)) \subset g(\mathbf{C}f(\mathbf{C}B)) = \varphi(B);$$

that is, φ satisfies the condition of 9.5. Hence, there is a set $D \in \mathscr{P}(X)$ such that $\varphi(D) = D$, that is

9.9
$$D = g(\mathbf{C}f(\mathbf{C}D)).$$

Note that 9.9 shows that $g(Y) \supset D$. Thus, for each $x \in D$, there is a unique $y \in Y$ satisfying $g(y) = x$, hence $y = g^{-1}(x)$.

Define now $\psi: X \to Y$ by

$$\psi(x) = g^{-1}(x) \quad \text{if} \quad x \in D,$$
$$\psi(x) = f(x) \quad \text{if} \quad x \in \mathbf{C}D.$$

By 9.9,

$$\psi(D) = g^{-1}(D) = \mathbf{C}f(\mathbf{C}D)$$

and

$$\psi(\mathbf{C}D) = f(\mathbf{C}D).$$

Hence

$$\psi(X) = \psi(D) \cup \psi(\mathbf{C}D) = \mathbf{C}f(\mathbf{C}D) \cup f(\mathbf{C}D) = Y;$$

that is, ψ *is surjective.*

Now let $x' \in X$, $x'' \in X$, $x' \neq x''$. If x' and x'' belong to D, then $\psi(x') \neq \psi(x'')$, since g is injective; if x' and x'' belong to $\mathbf{C}D$, then

$\psi(x') \neq \psi(x'')$, since f is injective. Now suppose $x' \in D$ and $x'' \in \mathbf{C}D$. Then $\psi(x') \in \psi(D) = \mathbf{C}f(\mathbf{C}D)$ and $\psi(x'') \in \psi(\mathbf{C}D) = f(D)$; hence $\psi(x') \neq \psi(x'')$. Therefore, ψ is *injective*.

We conclude that ψ is bijective, and hence that the theorem is proved.   ▲

9.10   Corollary.—*Every infinite part of $\mathbf{Z}$ is equipotent to $\mathbf{Z}$.*

Proof.—Let $A \subseteq \mathbf{Z}$ be an infinite part. Then the mapping $f : A \to \mathbf{Z}$ defined by $f(x) = x$ for $x \in A$ is an injection. By 9.3, A contains a countable set A_0. Hence, there is a bijection $g' : \mathbf{Z} \to A_0$. If $g : \mathbf{Z} \to A$ is the mapping $x \mapsto g'(x)$ of $\mathbf{Z}$ into A, then g is an injection. By Theorem 9.4, $\mathbf{Z}$ and A are equipotent.

9.11   Corollary.—*A subset of a countable set is finite or countable.*

Proof.—Let X be a countable set and let $A \subseteq X$. Suppose A is not finite. Since X is countable, there is a bijection $f : X \to \mathbf{Z}$. Now $f(A)$ is equipotent to A and $f(A) \subseteq \mathbf{Z}$. By 9.10, $f(A)$ is countable; hence A is countable.

9.12   Theorem.—*If X is countable and $f : X \to Y$ is a surjection, then Y is finite or countable.*

Proof.—For each $y \in Y$, $f^{-1}(y) \neq \varnothing$, since f is a surjection; then let $g(y) \in f^{-1}(y)$. Consider the mapping $y \mapsto g(y)$ of Y into X. If $y' \in Y$, $y'' \in Y$ and $y' \neq y''$, then $f^{-1}(y') \cap f^{-1}(y'') = \varnothing$; since $g(y') \in f^{-1}(y')$ and $g(y'') \in f^{-1}(y'')$ we deduce $g(y') \neq g(y'')$. Hence, g is an injection, and hence Y is equipotent to $g(Y)$. Since $g(Y) \subseteq X$ and X is countable, we deduce that $g(Y)$ is finite or countable (see 9.11). Hence Y is finite or countable.

Example 7.—*The set $N \times N$ is countable.*

▼   Let $f : N \times N \to N$ be defined by*
$$f(x, y) = \tfrac{1}{2}(x + y)(x + y + 1) + y$$

for $(x, y) \in N \times N$. We shall show that f is an *injection*. Let $(x, y) \in N \times N$ and $(x', y') \in N \times N$ be such that $(x, y) \neq (x', y')$. We shall distinguish three cases:

---

* At least one of the numbers $x + y$ or $x + y + 1$ is even, whence $\tfrac{1}{2}(x + y)(x + y + 1) \in N$.

Case I: $x + y = x' + y'$. Then $y \neq y'$ (otherwise, $x = x'$ too). We deduce

$$f(x,y) - f(x',y') = y - y' \neq 0$$

that is, $f(x,y) \neq f(x',y')$.

Case II: $x + y < x' + y'$. Since $x + y$ and $x' + y'$ belong to· N, we have $x + y + 1 \leq x' + y'$. Hence:

$$
\begin{aligned}
f(x,y) &= \tfrac{1}{2}(x+y)(x+y+1) + y \leq \tfrac{1}{2}(x+y)(x'+y') + y \\
&= \tfrac{1}{2}(x+y)(x'+y'+1) - \tfrac{1}{2}(x+y) + y \\
&\leq \tfrac{1}{2}(x'+y')(x'+y'+1) \\
&\quad - \tfrac{1}{2}(x'+y'+1) - \tfrac{1}{2}(x+y) + y \\
&\leq \tfrac{1}{2}(x'+y')(x'+y'+1) - (x+y) + y \\
&= \tfrac{1}{2}(x'+y')(x'+y'+1) - x \\
&< \tfrac{1}{2}(x'+y')(x'+y'+1) + y' \\
&= f(x',y');
\end{aligned}
$$

and hence, $f(x,y) \neq f(x',y')$.

Case III: $x + y > x' + y'$. Reasoning as in Case II, we show that $f(x,y) \neq f(x',y')$.

We deduce that f is *injective*. Therefore, $N \times N$ is equipotent to $f(N \times N) \subset N$. By 9.11, $f(N \times N)$ is finite or countable. Since $f(N \times N)$ contains a set equipotent to $\{(n,1) \mid n \in N\}$ (which is clearly equipotent to N), it follows that $f(N \times N)$ is infinite (using 9.1). Hence, $N \times N$ is countable.   ▲

Exercise 1.—Show that if $A_1, A_2, \ldots, A_n$ are countable sets, then $A_1 \times A_2 \times \ldots \times A_n$ is countable.

Exercise 2.—Show that if $(A_i)_{i \in N}$ is a family of countable sets, then $\bigcup_{i \in N} A_i$ is countable.

Example 8.— *The set Q of rational numbers is countable.*

The set $Z \times N$ is equipotent to $N \times N$ (see Example 6) and $N \times N$ is countable. The mapping

$$(n, m) \mapsto n/m$$

of $Z \times N$ into Q is a surjection; hence (using 9.12) the set Q is finite or countable. Since $Q \supset N$, we deduce that Q is countable.

It will be shown below that there are infinite sets that are not countable.

For every $a \in R$, $b \in R$, $a < b$ we write

9.13 $[a, b] = \{x \mid a \leq x \leq b\}.$

A set of this form is called a (bounded) *closed interval* (see Appendix I).

9.14   Theorem.—*Let $I \subset R$ be a closed interval. Then the set I is infinite and not countable.*

For the proof, we note that given a closed interval $J \subset R$ and $x \in R$, there exists a closed interval J' such that

$$x \notin J' \quad \text{and} \quad J' \subset J.$$

Proof.—By I.3 of Appendix I, the set I is infinite. If I were countable, there would be a bijection $n \mapsto x_n$ of N onto I. Let I_1 be a closed interval such that

$$x_1 \notin I_1 \quad \text{and} \quad I_1 \subset I.$$

Let I_2 be a closed interval such that

$$x_2 \notin I_2 \quad \text{and} \quad I_2 \subset I_1.$$

Continuing in this way,* we obtain a sequence $(I_n)_{n \in N}$ of closed intervals such that (denote I by I_0)

9.15 $x_n \notin I_n \quad \text{and} \quad I_n \subset I_{n-1} \quad \text{for all} \quad n \in N.$

By I.4 (Appendix I), there is $t \in \bigcap_{n \in N} I_n$, and hence (since clearly $t \in I$) there is $k \in N$ such that $t = x_k$. But $x_k \notin I_k$, thus

$$t \notin \bigcap_{n \in N} I_n.$$

Hence, the hypothesis that I was countable led to a contradiction; therefore I is not countable.

9.16   Corollary.—The set R of real numbers is infinite and not countable.

Proof.—By 9.2, R is infinite. By Corollary 9.11, it is not countable.

Therefore, $Z \subset R$, and both sets are infinite, although they are not equipotent. For a long time, mathematicians tried to establish the *continuum hypothesis*, that is, to establish that every set such that

---

* See Chapter 12.

$Z \subset A \subset R$ is either equipotent to Z or equipotent to R. Recently, it has been shown (see [2] and [3]) that neither the continuum hypothesis nor its negation can be proved on the basis of the usual axioms of set theory. Actually, if either the continuum hypothesis or its negation is added as an axiom, the theory obtained remains "consistent" (if the previous one is).

Exercises for Chapter 9

1. Let $I = \{1, 2, 3, 4, 5\}$, $X = \{0, 1\}$, $Y = X^5$. Define a mapping φ by

$$\varphi = (Y, \mathscr{P}(I), \{(x, \{i \mid pr_i(x) = 1\}) \mid x \in Y\}).$$

Is φ a bijection?

2. Show that $\mathscr{P}(\mathscr{P}(\mathscr{P}(\varnothing)))$ is equipotent to A_4 (see Exercise 3, Chap. 1).

3. Let A, B, and C be three sets such that $B \cap C = \varnothing$. Show that $A^{B \cup C}$ is equipotent to $A^B \times A^C$.

4. Show that the set $S = \{n^2 \mid n \in N\}$ is infinite. Show that S is countable.

Chapter 10

Relations in a Set

Let X be a set. A *relation in X* is a subset

$$R \subset X \times X.$$

Given $x \in X$ and $y \in X$, we say that x is R-related to y if $(x, y) \in R$.

Example 1.—Let X be a set and let $\Delta = \{(x, x) \mid x \in X\}$. Then Δ is a relation in X. Clearly

$$(x, y) \in \Delta \Leftrightarrow x = y.$$

Example 2.—Let X be a set, Y be a set, and $\varphi: X \to Y$ be a function. Let

$$R = \{(x, y) \mid x \in X, y \in X, \varphi(x) = \varphi(y)\}.$$

Then R is a relation in X. We emphasize that

$$(x, y) \in R \Leftrightarrow \varphi(x) = \varphi(y).$$

Example 3.—Let X be a set and $f: X \to X$. Then G_f (= the graph of f) is a relation in X. Note that

$$(x, y) \in G_f \Leftrightarrow y = f(x)$$

Example 4.—Let

$$R = \{(x, y) \mid x \in \mathbf{R}, y \in \mathbf{R}, y - x \in \mathbf{R}_+\}.$$

Then R is a relation in $\mathbf{R}$. Note that

$$(x, y) \in R \Leftrightarrow y - x \in \mathbf{R}_+.$$

Let X be a set and let R be a relation in X. Then we say that:

10.1 R is *reflexive* if $(x, x) \in R$ for all $x \in X$;

10.2 R is *symmetric* if $(x, y) \in R$ implies $(y, x) \in R$;

10.3 R is *transitive* if $(x, y) \in R$ and $(y, z) \in R$ implies $(x, z) \in R$.

Let X be a set and R a relation in X. We say that R is an *equivalence relation in X* if R is *reflexive, symmetric,* and *transitive.* Since equivalence relations are important in mathematics, we shall study them here in some detail.

If R is an equivalence relation in X, we often write

$$x \equiv y \ (\text{mod } R)$$

(read, "x equivalent to y modulo R") instead of $(x, y) \in R$. We also write $x \not\equiv y \ (\text{mod } R)$ instead of $(x, y) \notin R$.

Example 5.—The relations in Examples 1 and 2 are equivalence relations. As we shall see below, for every set X and equivalence relation R in X, there is a set Y and a function $\varphi : X \to Y$ such that

$$(x, y) \in R \Leftrightarrow \varphi(x) = \varphi(y).$$

Example 6.—Let $p \in \mathbf{Z}$ and let R_p be the set of all $(x, y) \in \mathbf{Z} \times \mathbf{Z}$ such that

$$x - y = np \text{ for some } n \in \mathbf{Z}.$$

Then R_p is an equivalence relation in $\mathbf{Z}$.

Let $x \in \mathbf{Z}$. Then $x - x = 0 \cdot p$; hence $(x, x) \in R_p$; thus R_p is *reflexive.* Now let $(x, y) \in R_p$. Then $x - y = np$ for some $n \in \mathbf{Z}$ and thus

$$y - x = -(x - y) = -(np) = (-n)p;$$

hence $(y, x) \in R_p$; thus R_p is *symmetric.* Finally, let $(x, y) \in R_p$ and $(y, z) \in R_p$. Then

$$x - y = np \text{ and } y - z = mp$$

for some $n \in \mathbf{Z}$, $m \in \mathbf{Z}$; whence $x - z = (n + m)p$; that is, $(x, z) \in R_p$, so R_p is *transitive.* We conclude that R_p *is an equivalence relation in Z.*

Example 7.—Let $X = \mathbf{Z} \times \mathbf{N}$, and let R be the set of all pairs $((p, q), (n, m)) \in X \times X$ such that

$$pm = qn.$$

Then R is an equivalence relation in $\mathbf{Z} \times \mathbf{N}$. Note that

$$(p, q) \equiv (n, m)(\text{mod } R) \Leftrightarrow \frac{p}{q} = \frac{n}{m}.$$

Example 8.—Let X be a set, $A \subset X$, and

$$R = (A \times A) \cup (\mathbf{C}A \times \mathbf{C}A).$$

Then R is an equivalence relation in X.

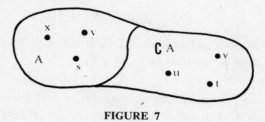

FIGURE 7

If x, y, u, v, s, t are as in Figure 7, then

$$x \equiv y \ (\text{mod } R), \ u \equiv v \ (\text{mod } R), \ s \not\equiv t \ (\text{mod } R).$$

Hence, two elements, x and y, are equivalent modulo R if they are both either in A or in $\mathbf{C}A$.

Example 9.—Let X be a set and let R be the set of all couples (A, B) such that:

(i) $A \in \mathscr{P}(X), B \in \mathscr{P}(X)$;
(ii) A and B are equipotent.

Then R is an equivalence relation in $\mathscr{P}(X)$.

Let X be a set and R an *equivalence relation* in X. For each $x \in X$, let

$$C(x) = \{y \,|\, y \in X, y \equiv x \ (\text{mod } R)\}.$$

Then $C(x) \in \mathscr{P}(X)$. The set $C(x)$ is called the *equivalence class* of $x \in X$.

The basic properties of equivalence classes are given in Theorem 10.4.

10.4   Theorem.—*Let $a \in X$ and $b \in X$.   Then the following assertions are equivalent:*

> (i) $a \equiv b \pmod{R}$;
> (ii) $C(a) = C(b)$;
> (iii) $C(a) \cap C(b) \neq \varnothing$.

Proof of 10.4(i) $\Rightarrow$ 10.4(ii).—Let $x \in C(a)$. Then $x \equiv a \bmod (R)$. Since $a \equiv b \pmod{R}$, and since R is an equivalence relation we, deduce $x \equiv b \pmod{R}$, whence $x \in C(b)$. Since $x \in C(a)$ was arbitrary, we deduce $C(a) \subset C(b)$. In the same way, we show that $C(b) \subset C(a)$. We conclude that $C(a) = C(b)$.

Proof of 10.4(ii) $\Rightarrow$ 10.4(iii).—Obvious, since $C(a) = C(b) \ni b$.

Proof of 10.4(iii) $\Rightarrow$ 10.4(i).—Let $c \in C(a) \cap C(b)$. Then

$$a \equiv c \pmod{R} \quad \text{and} \quad c \equiv b \pmod{R},$$

whence $a \equiv b \pmod{R}$.

Hence, Theorem 10.4 is proved.

10.5   Theorem.—*Let R be an equivalence relation in a set X. There exists then a set $\mathscr{C}$ and surjection $\varphi: X \to \mathscr{C}$ such that*

$$(x, y) \in R \iff \varphi(x) = \varphi(y).$$

Proof.—Let $\mathscr{C}$ be the set of all $A \in \mathscr{P}(X)$ having the property: There exists $x \in X$ such that $A = C(x)$. Hence, $\mathscr{C}$ is the set of elements of the family $(C(x))_{x \in X}$.

Let $\varphi: X \to \mathscr{C}$ be the mapping defined by

$$\varphi(x) = C(x)$$

for $x \in X$. Clearly, by its definition, $\varphi: X \to \mathscr{C}$ is a surjection. Furthermore, if

$(x, y) \in R,$ then $C(x) = C(y)$ that is, $\varphi(x) = \varphi(y).$

Conversely:

$\varphi(x) = \varphi(y)$ implies $C(x) = C(y)$; that is, $(x, y) \in R.$

Hence

$$(x, y) \in R \Leftrightarrow \varphi(x) = \varphi(y)$$

and hence Theorem 10.5 is proved.

The set $\mathscr{C}$ constructed in Theorem 10.5 is usually denoted by X/R and is called the *quotient set of X by R*. The mapping $\varphi : X \to X/R$ is called the *canonical mapping* associated with R. As we have seen, φ is a *surjection*.

If $x \in X$, then $\varphi^{-1}(\varphi(x)) = C(x)$ (here $C(x)$ is considered as a subset of X).

Exercises for Chapter 10

1. Let $X = \{1, 2, 3, 4\}$. Determine whether the following relations in X are reflexive, symmetric, or transitive.

    (a) $\{(1, 2), (2, 1), (1, 1)\} = A$;

    (b) $\{(1, 2), (2, 1), (1, 1), (4, 3)\} = B$;

    (c) $\{(1, 2), (3, 4), (1, 1), (2, 2), (3, 3), (4, 4)\} = C$;

    (d) $X \times X$;

    (e) $\varnothing$;

    (f) $A \cup B \cup C$.

2. Let X, A, B, and C be as in Exercise 1. Determine whether the following relations are reflexive, symmetric or transitive, (see Exercise 5, Chap. 3)

    (a) $A \circ B$;

    (b) $A \circ C$;

    (c) $B \circ A$;

    (d) $C \circ A$;

    (e) $B \circ C$;

    (f) $(X \times X) \circ A$.

3. Let X be a set and let A and B be relations in X.

    (a) Suppose A and B are reflexive. Are $A \cup B$ and $A \cap B$ reflexive?

    (b) Suppose A and B are symmetric. Are $A \cup B$ and $A \cap B$ symmetric?

    (c) Suppose A and B are transitive. Are $A \cup B$ and $A \cap B$ transitive?

Chapter II

Order Relations

In this chapter, we introduce the notion of order relation. We also introduce the notion of inductive set and state Zorn's lemma, which has many applications. We close with a brief discussion of cardinal numbers.

Let X be a set and let R be a relation in X. Then

11.1 R is *anti-symmetric* if

$$(x, y) \in R \quad \text{and} \quad (y, x) \in R \Rightarrow x = y.$$

Again, let X be a set and R a relation in X. We say that R is an *order relation* in X if R is reflexive, anti-symmetric, and transitive.

If R is an order relation in X, we often write $x \leq y$ or $y \geq x$ instead of $(x, y) \in R$. We also write $x < y$ or $y > x$ instead of $(x, y) \in R$ and $x \neq y$.

With the notation just given, we may formulate the definition of an order relation in the following way:

Let X be a set and R a relation in X. Then R is an order relation in X if and only if:

(1) $x \leq x$ for all $x \in X$;
(2) $x \leq y$ and $y \leq x \Rightarrow x = y$;
(3) $x \leq y$ and $y \leq z \Rightarrow x \leq z$.

We shall often use the expression, "Let X be a set endowed with an order relation." Whenever we use this expression, we consider a certain given order relation in X and we fix our attention on the set X and on the considered order relation in X. By an *ordered set* we mean a set endowed with an order relation.

Example 1.—The relation in Example 4, Chapter 10 is an order relation in R. This is the order relation R is usually endowed with.

Let X be a set, R an order relation in X, and x, y, and z three elements of X. It is then easy to see that the following hold:

11.2 $$x \leq y \quad \text{and} \quad y < z \Rightarrow x < z;$$

11.3 $$x < y \quad \text{and} \quad y \leq z \Rightarrow x < z;$$

11.4 $$x < y \quad \text{and} \quad y < z \Rightarrow x < z.$$

Again, let X be a set and R an order relation in X. Let $A \subset X$; then $b \in X$ is a *majorant* of A if $b \geq x$ for all $x \in A$. A set $A \subset X$ is said to be *bounded above* if it has a majorant.

Now let $A \subset X$, $A \neq \varnothing$. We say that β is a *supremum* of A if:

11.5 $$\beta \text{ is a majorant of } A;$$

11.6 $$\text{if } b \text{ is a majorant of } A, \text{ then } b \geq \beta.$$

A set A (even bounded above) does not have necessarily a supremum.

Notice that if β_1 and β_2 are supremums of A, then by 11.6 above we have $\beta_1 \leq \beta_2$ and $\beta_2 \leq \beta_1$; whence by (2) $\beta_1 = \beta_2$. Hence a set A has *at most one* supremum. The supremum of a set A will be denoted (when it exists) by

11.7 $$\sup A.$$

Again, let $A \subset X$; then, $a \in X$ is a *minorant* of A if $a \leq x$ for all $x \in A$. A set A is said to be *bounded below* if it has a minorant.

Now let $A \subset X$, $A \neq \varnothing$. We say that α is an *infimum* of A if:

11.8 $$\alpha \text{ is a minorant of } A;$$

11.9 $$\text{if } a \text{ is a minorant of } A, \text{ then } a \leq \alpha.$$

As in the case of supremums, it follows immediately that a set A has at most one infimum. It might have none, even if it is bounded below. The infimum of a set A will be denoted (when it exists) by

11.10 $$\inf A.$$

A set $A \subset X$ is said to be *bounded* if it has both a majorant and a minorant.

A relation R in a set X is said to be a *total order relation* if it is an order relation and if it satisfies the following condition:

(4) We have either $x \leq y$ or $y \leq x$ for all $x \in X$ and $y \in X$.

A *totally ordered set* is a set endowed with a total order relation.

Remark.—The order relation in $\mathbf{R}$ mentioned in Example 1 is a total order relation.

Example 2.—Let Y be a set and let

$$R = \{(A, B) \mid A \subset B \subset Y\}.$$

Then R is an order relation in $\mathscr{P}(Y)$. Moreover, R is a total order relation if and only if either $X = \varnothing$ or X contains only one element.

A relation R in a set X is said to be a *well-order relation* if it is an order relation and if it satisfies the following condition:

(5) Every non-void part of X contains a smallest element.

This means that if $A \neq \varnothing$ is a part of X, then there exists (a smallest element) a such that $a \in A$ and $a \leq x$ for all $x \in A$.

A *well-ordered set* is a set endowed with a well-order relation.

Example 3.—The set $\mathbf{N}$ endowed with the order relation

$$R = \{(x, y) \mid x \in \mathbf{N}, y \in \mathbf{N}, x - y \in \mathbf{Z}_+\}$$

is well-ordered.

Example 4.—The set $\mathbf{R}$ (endowed with the usual order relation) is not well-ordered. In fact, let

$$\mathbf{R}_+^* = \{x \mid x \in \mathbf{R}, x > 0\}.$$

If $a \in \mathbf{R}_+^*$, then $a/2 \in \mathbf{R}_+^*$ and $a/2 < a$. We conclude that $\mathbf{R}_+^*$ does not contain a smallest element.

It is easy to see that a well-ordered set is totally ordered. In fact, let $x \in X$, $y \in X$, and $A = \{x, y\}$. Then $A \subset X$, and hence it contains a smallest element; hence either $x \leq y$ or $y \leq x$. Hence, X is totally ordered. That a totally ordered set is not necessarily well-ordered was shown in Example 4.

A set X is said to be *directed* (or *filtering*) if every non-void finite part of X is bounded above.

A set X is said to be a *lattice* if every non-void finite part has an infimum and a supremum.

Clearly, a totally ordered set is a lattice and a lattice is filtering.

Before proceeding further, we shall indicate a simple method of giving examples of finite ordered sets by certain drawings (Figs. 8–13).

The elements of such finite sets are represented by circles, while a pair (x, y) is a member of the considered order relation if and only if $x = y$ or x can be "connected" to y by a descending sequence of line segments. For example, the order relation R corresponding to Figure 8 is $\Delta \cup \Delta'$ where

$$\Delta = \{(1, 1), (2, 2), (3, 3), (4, 4), (5, 5)\}$$

and

$$\Delta' = \{(1, 3), (1, 5), (1, 4), (2, 3), (2, 5), (2, 4), (3, 4), (3, 5)\}.$$

Figure 10 represents a lattice. Figure 12 represents a totally ordered set. Figure 9 represents a directed set having a supremum but not an infimum; this is not a lattice. Figure 11 represents a set having an infimum but not a supremum. The subset $\{3, 4\}$ of the ordered set represented by Figure 13 is bounded above but has no supremum.

Let X be a set and R an order relation on X. If $A \subset X$, then clearly

$$R_A = (A \times A) \cap R$$

is an order relation on A. It is called the order relation induced on A by R. Whenever we consider A as a subset of X, we suppose (unless we mention explicitly the contrary) that A is endowed with the order relation R_A.

It follows immediately that if X is totally ordered, then A is totally ordered; if X is well-ordered, then A is well-ordered.

Let X be an ordered set. We say that $A \subset X$ is a *cofinal part* of X if for every $x \in X$ there exists $a_x \in A$ such that $a_x \geq x$. If X is a directed set and $A \subset X$ is cofinal, then A is directed.

Exercise.—Show that if X is directed and $A \subset X$ is arbitrary, then A is not necessarily directed. Show that if X is a lattice and $A \subset X$ is arbitrary, then A is not necessarily a lattice.

Let X be an ordered set. Then:

11.11 *An element $\beta \in X$ is said to be maximal if there is no $y \in X$ such that $y > \beta$.*

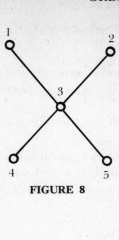

FIGURE 8

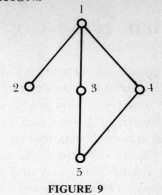

FIGURE 9

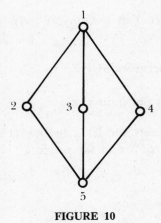

FIGURE 10

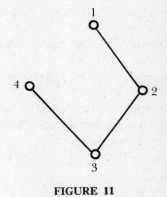

FIGURE 11

FIGURE 12

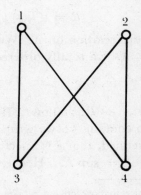

FIGURE 13

11.12   *An element* $\alpha \in X$ *is said to be minimal if there is no* $x \in X$
such that $x < \alpha$.

Hence, an element $\beta \in X$ is maximal if and only if $y \in X$ and
$y \geq \beta$ implies $y = \beta$. An element $\alpha \in X$ is minimal if and only if
$x \in X$ and $x \leq \alpha$ implies $x = \alpha$.

Clearly, if X has a supremum b, then b is a maximal element of
X and is the *only* maximal element of X.

The set in Figure 8 has neither supremum nor infimum. However, 1 and 2 are maximal elements and 3 and 4 are minimal
elements. The set in Figure 9 has a supremum, but has no infimum.
The elements 2 and 5 are minimal.

We shall now introduce the following:

11.13   Definition.—*An ordered set X is inductive if every totally
ordered subset of X has a majorant.**

Example 5.—Any finite ordered set is inductive.

Example 6.—The ordered set Z is not inductive.

Example 7.—Let S and T be two sets and let X be the set of all
injections having for domain a subset of S and for range T. For f
and g in X, we write

$$f \leq g$$

if:
  (i) $\operatorname{dom} f \subset \operatorname{dom} g$;
  (ii) $f(x) = g(x)$ for all $x \in \operatorname{dom} f$;
Hence, $f \leq g$ if and only if $G_f \subset G_g$. It is easy to show that

$$R = \{(f, g) \mid f \in X, g \in X, f \leq g\}$$

is an order relation on X. We shall now show that X is inductive.
Let $K \subset X$ be a totally ordered part. Let

$$A = \bigcup_{f \in K} \operatorname{dom} f.$$

Define $h : A \to T$ as follows: If $x \in A$, then there is $f \in K$ the domain
of which contains x; we write $h(x) = f(x)$. It is easy to see that h
is well defined, that h is injective, and that h is a majorant of K (in
fact, here $h = \sup K$). Hence, X is inductive.

---

* Notice that we do not suppose that every totally ordered subset of X has a
supremum.

We shall now state (without proof*) an important result that has many applications:

11.14   Zorn's Lemma.—*Every non-void inductive set has a maximal element.*

Some applications of Zorn's lemma will be indicated as we go along.

▼   Zorn's lemma is "equivalent" to the Axiom of Choice, which was stated near the end of Chapter 8. Hence, Zorn's lemma can be taken as an axiom. In any case, it is preferable (when possible) to take for axioms statements that intuitively are more readily acceptable, and the Axiom of Choice is clearly such a statement.

We would like, however, to take this opportunity to make the following remarks concerning the Axiom of Choice. Given a finite number of sets, it is "easy" to choose an element in each of them. But what about making the same choice in the case of an infinite family of sets? After long discussions (at the beginning of the century), it was finally recognized that the possibility of making such a choice is based on an axiom, namely, the Axiom of Choice.

We would like to mention here that the following assertion is "equivalent" with the Axiom of Choice and Zorn's lemma.

11.15   The Well-ordering Lemma.—*Every set X can be endowed with a well-order relation.*

For further details concerning these remarks, the reader may consult [10].   ▲

Consider the ordered set X in Example 7. Let $u \in X$. Then the following assertions are equivalent:

(i) u is maximal;
(ii) dom $u = S$ or u (dom u) $= T$.

Proof of (i) $\Rightarrow$ (ii).—Let u be maximal and suppose dom $u \neq S$ and u (dom u) $\neq T$. Let

$$s \in S - \text{dom } u \quad \text{and} \quad t \in T - u \text{ (dom } u).$$

Define $v: \{s\} \cup \text{dom } u \to T$ by:

$$v(x) = t \quad \text{if} \quad x = s;$$
$$v(x) = u(x) \quad \text{if} \quad x \in \text{dom } u.$$

_____

* The interested reader may consult, for instance, [1] or [6] .

Clearly, $v \in X$ and $v > u$. Since u is maximal, this leads to a contradiction; hence the implication (i) $\Rightarrow$ (ii) is proved.

Proof of (ii) $\Rightarrow$ (i).—Let $u \in X$ be such that dom $u = S$. If $h \geq u$, then dom $h \supset$ dom u, that is, dom $h = S$. We deduce $h = u$, and hence u is maximal. Now let $u \in X$ be such that $u(\text{dom } u) = T$. If $k \in X$ and $k \geq u$, then dom $k \supset$ dom u and $k(\text{dom } u) = T$. Hence

$$k(\text{dom } u) = u(\text{dom } u) = T.$$

Since k is an injection, we deduce dom $k =$ dom u. Hence $k = u$, and hence u is maximal. Hence, (ii) $\Rightarrow$ (i) is proved.

As a first application of Zorn's lemma we shall prove the following theorem.

11.16   Theorem.—*Let S and T be two sets. Then, one of the following assertions holds:*

11.17   *There is an injection of S into T;*

11.18   *There is an injection of T into S.*

Proof.—Let X be the set of all injections having for domain a subset S and range T, endowed with the order introduced in Example 7. Then X is inductive. By Zorn's lemma, there exists a maximal element $u \in X$. By the previous discussion, either dom $u = S$ or $u(\text{dom } u) = T$. If dom $u = S$, then u is an injection of S into T, and hence 11.17 holds. If $u(\text{dom } u) = T$, then $v: T \to S$, defined by

$$v(x) = u^{-1}(x) \quad \text{for} \quad x \in T$$

is an injection of T into S, and hence 11.18 holds. Hence, Theorem 11.16 is proved.

If there is an injection of S into T and an injection of T into S, then by the Bernstein-Schroeder theorem (see Theorem 9.4) *the sets S and T are equipotent.*

Hence, if S and T are two sets, then one of the following assertions holds:

11.19   *S and T are equipotent;*

11.20   *S is equipotent to a part of T;*

11.21   *T is equipotent to a part of S.*

Clearly, S and T are equipotent if and only if S is equipotent to a part of T and T is equipotent to a part of S.

In dealing with equipotent sets, the following theorem is sometimes useful.

11.22   Theorem.—*Let X be an infinite set.   Then there exists a partition* of X consisting of countable sets.*

▼   *Proof.*—Let $\mathscr{F}$ be the set of all parts $\mathscr{A} \subset \mathscr{P}(X)$ having the properties:

   (i) $A \in \mathscr{A}$ and $B \in \mathscr{A}$ and $A \neq B \Rightarrow A \cap B = \varnothing$.

   (ii) $A \in \mathscr{A} \Rightarrow A$ countable.

We shall "order $\mathscr{F}$ by inclusion"; hence, if $\mathscr{A}$ and $\mathscr{B}$ belong to $\mathscr{F}$, then $\mathscr{A} \leq \mathscr{B}$ if and only if $\mathscr{A} \subset \mathscr{B}$. It is easy to see that $\mathscr{F}$ is inductive when endowed with this order relation. Let $\mathscr{A}'$ be a maximal element of $\mathscr{A}$ and let

$$Z = X - \bigcup_{A \in \mathscr{A}'} A.$$

If $Z = \varnothing$, then the theorem is proved (notice that if $X_A = A$ for all $A \in \mathscr{A}'$, then $(X_A)_{A \in \mathscr{A}'}$ is a partition of X). If Z is *finite*, we pick $A \in \mathscr{A}'$, and we denote by $\mathscr{A}''$ the set obtained from $\mathscr{A}'$ by replacing the element A by $A \cup Z$. Clearly then, $\mathscr{A}'' \in \mathscr{F}$ and

$$\bigcup_{A \in \mathscr{A}'} A = X.$$

Hence, Theorem 11.22 is again proved. If Z is infinite, then there exists a countable set $Z_0 \subset Z$ (see 9.3). Since $\mathscr{A}''' = \mathscr{A}' \cup \{Z_0\} \in \mathscr{F}$ and $\mathscr{A}''' > \mathscr{A}'$, we arrive at a contradiction. Hence, the only possible cases are $Z = \varnothing$ or Z finite, and then the assertion in the theorem holds.   ▲

To each set A, an object, which we denote Card A (and call the *cardinal number* of A), can be associated such that:

Two sets X and Y are equipotent if and only if Card $X =$ Card Y.

If there were a set U such that every element of U is a set, and every set is an element of U, then we could consider the relation

$$R = \{(A, B) \mid A \in U, B \in U, A \text{ and } B \text{ are equipotent}\}.$$

---

* A *partition* of X is a family $(A_i)_{i \in I}$ of non-void parts of X such that $\bigcup_{i \in I} A_i = X$ and $A_i \cap A_j = \varnothing$ if $i \neq j$.

As we have seen in Example 9, Chapter 10, R is an equivalence relation in U. We could then have defined Card A, for each set A, as the equivalence class of A. However, as we indicated in Chapter 8, such a set U does not exist. Other methods are used to define Card A for a set A (the interested reader may consult, for instance, [1]).

In this volume, the numbers 0, 1, 2, . . . were supposed to be known. However, once the cardinal numbers are introduced, we may define 0, 1, 2, . . . as the cardinal numbers of finite sets. We could define

$$0 = \text{Card } \varnothing, \qquad 1 = \text{Card } \{\varnothing\}, \qquad 2 = \text{Card } \{\varnothing, \{\varnothing\}\}, \dots$$

Once the numbers 0, 1, 2, . . . are introduced, we can successively construct Z and Q, and then R (see, for instance, [4]).

Exercises for Chapter 11

1. Let X be an ordered set and $A \subset X$ a non-void set having both an infimum and a supremum. Then inf $A \le$ sup A. (If X has a smallest element m and a largest element M, and if $A = \varnothing$, we define inf $A = M$ and sup $A = m$.)

2. Let X be a set and R an order relation in X. Then R^{-1} $(= \{(x, y) \,|\, (y, x) \in R\})$ is an order relation in X. If $A \subset X$, then* $\sup_R A$ exists if and only if $\inf_{R^{-1}} A$ exists; if $\sup_R A$ exists then $\sup_R A = \inf_{R^{-1}} A$. Also, $\inf_R A$ exists if and only if $\sup_{R^{-1}} A$ exists; if $\inf_R A$ exists then $\inf_R A = \sup_{R^{-1}} A$.

3. Let X be an ordered set, I a set, and $f : I \to X$. We say that f has a supremum on $A \subset I$ if $f(A)$ has a supremum. If f has a supremum on A, we write sup $f(x) = \sup f(A)$.† Let $(A_j)_{j \in J}$ be
$\quad\quad\;\; x \in A$
a family of parts of I the union of which is I. Suppose that sup $f(x)$
$\quad\quad\quad\quad\quad\quad\quad\quad\quad\quad\quad\quad\quad\quad\quad\quad\quad\quad\quad\quad\quad x \in A_j$
exists for all j in J. Then f has a supremum on I if and only if

---

* Let Y be a set and S an order relation on Y and $A \subset Y$. If the supremum of A "with respect to S" exists, we may write $\sup_S A$ instead of sup A. Similarly, if the infimum of A "with respect to S" exists, we may write $\inf_S A$ instead of inf A.

† In a similar way we introduce inf $f(x)$.
$\quad\quad\quad\quad\quad\quad\quad\quad\quad\quad\quad\quad\quad\quad\quad x \in A$

$j \mapsto \sup\limits_{x \in A_j} f(x)$ has a supremum on J. If f has a supremum on I, then:

$$\sup_{x \in I} f(x) = \sup_{j \in J} \left(\sup_{x \in A_j} f(x) \right).$$

4. Discuss Exercise 3 in the case in which $I = X$ and $f = j_X$ (j_X is the identity mapping of X onto X).

5. Let X be a well-ordered set such that every non-void subset of X has a largest element. Show that X is finite.

Chapter 12

Mathematical
Induction

We shall often make use in this volume of the following, which we accept without proof.

12.1   Theorem (Mathematical Induction).—*Let $S \subset N$ be a set satisfying the following two conditions:*

12.2 $1 \in S;$

12.3 $n \in S \Rightarrow n + 1 \in S.$

Then $S = N$.

Note that N satisfies 12.2 and 12.3.

Proofs that make use of 12.1 are called *proofs by mathematical induction* or simply *proofs by induction*.

Example 1.—Let A be a set and let $f : N \to A$, $g : N \to A$ be two functions. Suppose that:

12.4 $f(1) = g(1);$

12.5 $n \in N$ and $f(n) = g(n) \Rightarrow f(n + 1) = g(n + 1).$

Then

12.6 $f = g.$

In fact, let
$$S = \{n \mid n \in N, f(u) = g(u)\}.$$

By 12.4, $S \ni 1$. If $n \in S$, then $f(n) = g(n)$; by 12.5 we deduce $f(n + 1) = g(n + 1)$. Hence, $n \in S \Rightarrow n + 1 \in S$. Hence, S satisfies

12.2 and 12.3, and hence $S = N$. Therefore, $f(n) = g(n)$ for all $n \in N$; that is, $f = g$.

Example 2.—We have

12.7 $$1 + 2 + \ldots + n = \frac{n(n + 1)}{2}$$

for all $n \in N$.

Let f and g (on N to N) be defined by

$$f(n) = 1 + 2 + \ldots + n \quad \text{and} \quad g(n) = \frac{n(n + 1)}{2}$$

for $n \in N$. Clearly, $f(1) = g(1)$. Suppose now that $n \in N$ and $f(n) = g(n)$; then

$$f(n + 1) = f(n) + (n + 1) = g(n) + (n + 1)$$
$$= \frac{n(n + 1) + 2(n + 1)}{2} = \frac{(n + 1)(n + 2)}{2}$$
$$= g(n + 1).$$

Hence, f and g satisfy 12.4 and 12.5, whence $f = g$. We conclude that 12.7 holds for all $n \in N$.

Example 3.—We have

12.8 $$2^n > n$$

for all $n \in N$.

Let
$$S = \{n \mid n \in N, 2^n > n\}.$$

Clearly, $2^1 = 2 > 1$, whence $1 \in S$. Suppose further that $n \in S$. Then $2^n > n$; hence

$$2^{n+1} = 2 \cdot 2^n > 2 \cdot n = n + n \geq n + 1.$$

Therefore, $2^{n+1} > n + 1$; that is, $n + 1 \in S$. By *mathematical induction*, we conclude that $S = N$; that is, that 12.8 holds for all $n \in N$.

Example 4.—If $x \in R$, $x > -1$, and $n \in N$, then

12.9 $$(1 + x)^n \geq 1 + nx.$$

Inequality 12.9 is called *Bernoulli's inequality.*

Let
$$S = \{n \mid n \in N, (1 + x)^n \geq 1 + nx\}.$$

Clearly, $1 \in S$, since $(1 + x)^1 = 1 + x = 1 + 1 \cdot x$. Suppose now that $n \in S$. Then

$$(1 + x)^{n+1} = (1 + x)^n (1 + x) \geq (1 + nx)(1 + x)$$
$$= 1 + nx + x + nx^2 \geq 1 + nx + x$$
$$= 1 + (n + 1)x.$$

Hence $n + 1 \in S$. By *mathematical induction* we conclude that $S = N$, that is, that 12.9 holds for all $n \in N$.

Exercise.—For every $n \in N$, we have

$$1^2 + 2^2 + \ldots + n^2 = \frac{n(n + 1)(2n + 1)}{6}.$$

(Hint: Use mathematical induction.)

Exercise.—If $x \in R$, $x \neq -1$, then for every $n \in N$ we have

$$1 + x + \ldots + x^{n-1} = (1 - x^n)/(1 - x).$$

(Hint: Use mathematical induction.)

Exercise.—Let A be a finite set and let $p = c(A)$. Show that $c(\mathscr{P}(A)) = 2^p$.
(Hint: Use mathematical induction.)

An important consequence of the induction theorem is the following theorem.

12.10 The Recursion Theorem.—*Let X be a set and $a \in X$. Suppose that $f : X \to X$. Then there is a function $F : N \to X$ such that:*

(i) $F(1) = a$;
(ii) $F(n + 1) = f(F(n))$ *for every $n \in N$.*

▼ *Proof.*—Let $\mathscr{A}$ be the set of all $A \subseteq N \times X$ having the properties:

12.11 $\qquad\qquad\qquad (1, a) \in A$;

12.12 $\qquad\quad (n, x) \in A \Rightarrow (n + 1, f(x)) \in A.$

We have $N \times X \in \mathscr{A}$; hence $\mathscr{A} \neq \varnothing$. Now let $E = \bigcap \mathscr{A}$. We will show that (N, X, E) is a function; i.e., that $pr_1(E) = N$, and that $(p, x) \in E$ and $(p, y) \in E \Rightarrow x = y$. Let

$$S = \{n \mid (n, x) \in E \text{ for one and only one } x\}.$$

We know that $(1, a) \in E$. Suppose there were an element $b \neq a$ such that $(1, b) \in E$. Consider the set $E_1 = E - \{(1, b)\}$. We have $(1, a) \in E_1$. If $(n, x) \in E_1$, then $(n + 1, f(x)) \in E$; hence $(n + 1, f(x)) \in E_1$ (because $n + 1 \neq 1$). This shows that $E_1 \in \mathscr{A}$. Since this contradicts the definition of E, we deduce that $1 \in S$.

Assume now that $n \in S$. Then there is a unique $x \in X$ such that $(n, x) \in E$. By 12.12, $(n + 1, f(x)) \in E$. Suppose that there is $y \in X$ such that $y \neq f(x)$ and $(n + 1, y) \in E$. Consider the set $E_n = E - \{(n + 1, y)\}$. We have $(1, a) \in E_n$. Suppose $(p, z) \in E_n$. If $p \neq n$, then $p + 1 \neq n + 1$; hence $(p + 1, f(z)) \in E_n$. If $p = n$, then (by the definition of S) $z = x$; therefore $f(z) = f(x) \neq y$ and $(p + 1, f(z)) \in E_n$. This shows that $E_n \in \mathscr{A}$. Since this contradicts the definition of E, we deduce that $n + 1 \in S$ whenever $n \in S$.

Thus, $S = N$ by Theorem 12.1. We conclude that $pr_1 E = N$ and that (N, X, E) is a function. Denote this function by F. Let $n \in N$ be arbitrary. Since $(n, F(n)) \in E$, we have $(n + 1, f(F(n))) \in E$; that is, $F(n + 1) = f(F(n))$. Since, obviously, $F(1) = a$, the theorem is proved.

Theorem 12.10 is often used, although not always explicitly. For instance, it can be used to justify the construction of the sequence $(I_n)_{n \in N}$ in Theorem 9.14. We proceed as follows: Let $X = N \times \mathscr{I}$, when $\mathscr{I}$ is the set of all closed intervals contained in I. For each $(n, J) \in X$, let $f(n, J) = (n + 1, J')$ where $J' \in \mathscr{I}$, $J' \subset J$, and $x_n \notin J'$. We define in this way a function $f: X \to X$. By Theorem 12.10, there is $F: N \to X$ such that $F(1) = (1, I)$ and $F(n + 1) = f(F(n))$ for every $n \in N$. If we define $I_{n-1} = pr_2(F(n))$ for every $n \in N$, then $(I_n)_{n \in N}$ satisfies the conditions in Theorem 9.14. ▲

Exercises for Chapter 12

1. Show that $1^2 + 2^2 + 3^2 + \ldots + n^2 = \dfrac{n(n + 1)(2n + 1)}{6}$ for $n \in N$.

2. Show that $1 + 3 + \ldots + (2n - 1) = n^2$ for $n \in N$.

3. Show that $1^3 + 2^3 + 3^3 + \ldots n^3 = \dfrac{n^2(n + 1)^2}{4}$ for $n \in N$.

4. Show that for each $n \in N$, there is $n' \in N$ such that $(4^n - 1) = 3n'$.

Chapter 13

Combinatorial Analysis

Let A be a finite set having p elements. Then there exists a bijection $\varphi : \{1, \ldots, p\} \to A$. If we write $a_j = \varphi(j)$ for $1 \le j \le p$, then

$$A = \{a_1, a_2, \ldots, a_p\}.$$

(note that here $a_i \ne a_j$ if $i \ne j$). Recall that if A is a finite set, then we denote by $c(A)$ the number of elements of A. Recall that (see 9.1) if A is a finite set and $B \subset A$, then B and $A - B$ are finite and

13.1 $$c(A) = c(B) + c(A - B).$$

Let A and B be two sets and let $f : A \to B$ be an injection. Then

$$A \text{ is finite} \Leftrightarrow f(A) \text{ is finite}.$$

Moreover, if A and $f(A)$ are finite, they have the same number of elements. It also follows that A is finite if B is.

If A and B are finite sets having the same number of elements, and $f : A \to B$ is an *injection*, then f is a *bijection*. In fact, as we have seen, A and $f(A)$ have the same number of elements; that is, $c(A) = c(f(A))$. Now

$$\begin{aligned}
c(B) &= c(f(A)) + c(B - f(A)) \\
&= c(A) + c(B - f(A)) \\
&= c(B) + c(B - f(A));
\end{aligned}$$

hence $c(B - f(A)) = 0$; that is, $B - f(A) = \varnothing$. We deduce that $f(A) = B$. Hence f is a surjection, and hence f is a bijection.

We can generalize 13.1 as follows: Let A be a finite set and let $A_1, A_2, \ldots, A_n (n > 1)$ be pairwise disjoint sets whose union is A. Then

13.2 $\qquad c(A) = c(A_1) + c(A_2) + \ldots + c(A_n).$

▼ The assertion 13.2 can be proved by mathematical induction. Let S be the set of all $m \in N$ such that if $B_1, B_2, \ldots, B_{m+1}$ are disjoint parts of A whose union is A, we have

$$c(A) = c(B_1) + c(B_2) + \ldots + c(B_{m+1}).$$

From 13.1, we deduce that $1 \in S$ (note that $B_1 \cap B_2 = \varnothing$, $B_1 \cup B_2 = A \Leftrightarrow B_1 \subset A$, and $B_2 = A - B_1$). Suppose now that $m \in S$. Let $D_1, D_2, \ldots, D_{m+2}$ be pairwise disjoint parts of A whose union is A. Then

$$(D_1 \cup D_2 \cup \ldots \cup D_{m+1}) \cap D_{m+2} = \varnothing$$

and

$c(D_1 \cup D_2 \cup \ldots \cup D_{m+1} \cup D_{m+2})$
$$= c(D_1 \cup D_2 \cup \ldots \cup D_{m+1}) + c(D_{m+2}).$$

Since $m \in S$,

$$c(D_1 \cup D_2 \cup \ldots \cup D_{m+1}) = c(D_1) + c(D_2) + \ldots + c(D_{m+1});$$

hence

$c(D_1 \cup D_2 \cup \ldots \cup D_{m+2})$
$$= c(D_1) + c(D_2) + \ldots + c(D_{m+1}) + c(D_{m+2}).$$

Since $D_1, D_2, \ldots, D_{m+2}$ were pairwise disjoint parts of A whose union is A, we deduce $m + 1 \in S$. By mathematical induction, we conclude $S = N$. Hence 13.2 is true if $A_1, A_2, \ldots, A_n (n > 1)$ are arbitrary pairwise disjoint parts of A, whose union is A. ▲

The reader should convince himself that: *If $A_1, \ldots, A_n$ are finite sets, then:*

13.3 $A_1 \cup \ldots \cup A_n$ *is finite;*

13.4 $A_1 \times \ldots \times A_n$ *is finite and*

$$c(A_1 \times \ldots \times A_n) = c(A_1) \times \ldots \times c(A_n).$$

Hint for the proof of 13.4: We notice that

$$A_1 \times A_2 = \bigcup_{x \in A_2} A_1 \times \{x\}$$

and that $(A_1 \times \{x\}) \cap (A_2 \times \{y\}) = \varnothing$ if $x \in A_2, y \in A_2$ and $x \neq y$.

13.5   Theorem.—*For every finite set X, the set $\mathscr{P}(X)$ is finite and*

$$c(\mathscr{P}(X)) = 2^{c(X)}.$$

▼   *Proof.*—The assertion of the theorem is obvious if $X = \varnothing$ or if $c(X) = 1$. In fact, in the first case, $\mathscr{P}(X) = \{\varnothing\}$, while in the second case $\mathscr{P}(X) = \{\varnothing, X\}$; whence $c(\mathscr{P}(X)) = 1 = 2^0$ if $X = \varnothing$ and $c(\mathscr{P}(X)) = 2 = 2^1$ if $c(X) = 1$.

To prove the theorem in general, we shall use mathematical induction. We denote by S the set of all $m \in N$ such that if X is a finite set having m elements, then $\mathscr{P}(X)$ is finite and $c(\mathscr{P}(X)) = 2^{c(X)}$. By the previous remarks, $1 \in X$. Now suppose that $m \in S$, and let X be a finite set having $m + 1$ elements. Let $a \in X$ and $X' = X - \{a\}$; then $c(X') = m$, and hence

$$c(\mathscr{P}(X')) = 2^m.$$

Note that $A \in \mathscr{P}(X)$ means that either $A \in \mathscr{P}(X')$ or $A \in \mathscr{P}'$, where $\mathscr{P}'$ is defined by

$$A \in \mathscr{P}' \Leftrightarrow A = B \cup \{a\} \quad \text{with} \quad B \in \mathscr{P}(X').$$

Now, $\mathscr{P}(X') \cap \mathscr{P}' = \varnothing$, (since $a \notin A$ if $A \in \mathscr{P}(X')$ and $a \in A$ if $A \in \mathscr{P}'$), also $\mathscr{P}'$ is finite and $c(\mathscr{P}') = 2^m$. Hence

$$c(\mathscr{P}(X)) = c(\mathscr{P}(X')) + c(\mathscr{P}') = 2^m + 2^m = 2 \cdot 2^m = 2^{m+1}. \quad ▲$$

13.6   Corollary.—*Let A and B be two finite sets. Then the set $\mathscr{F}(A, B)$ is finite.*

Recall that $\mathscr{F}(A, B)$ is the set of all mappings on A to B.

Proof.—Since $A \times B$ is finite, it follows from Theorem 13.5 that $\mathscr{P}(A \times B)$ is finite. By the result of Example 4 in Chapter 5, the mapping $f \to G_f$ is an injection of $\mathscr{F}(A, B)$ into $\mathscr{P}(A \times B)$. Since $\mathscr{P}(A \times B)$ is finite, we deduce that $\mathscr{F}(A, B)$ is finite.

13.7   Theorem*.—*Let X and Y be finite sets and $f : X \to Y$ a surjection. Suppose that $c(f^{-1}(y)) = a$ for some $a \in N$ and all $y \in Y$. Then*

$$c(X) = c(Y) \cdot a.$$

---

* This result is sometimes called "The shepherd's theorem."

Proof.—Let $Y = \{y_1, \ldots, y_n\}$ (with $y_i \neq y_j$ if $i \neq j$) and let

$$A_1 = f^{-1}(y_1), \ldots, A_n = f^{-1}(y_n).$$

Then (see 5.9 and 5.10) $A_k \cap A_j = \varnothing$ if $k \neq j$, and

$$A_1 \cup \ldots \cup A_n = X.$$

By 13.2, we conclude

$$c(X) = c(A_1) + \ldots + c(A_n) = n \cdot a = c(X) \cdot a.$$

Thus, Theorem 13.7 is proved.

For every *integer* $p \geq 0$ we define $p!$ by the equations:

$$0! = 1$$

$$1! = 1$$

$$p! = 1 \cdot 2 \ldots p \quad \text{if} \quad p > 1.$$

Thus $3! = 1 \cdot 2 \cdot 3 = 6$, $2!/0! = 2/1 = 2$, $5! = 1 \cdot 2 \cdot 3 \cdot 4 \cdot 5 = 120$.

For every pair of sets, X and Y, we denote by $\mathscr{I}(X, Y)$ the set of all injections of X into Y. Since $\mathscr{I}(X, Y) \subset \mathscr{F}(X, Y)$, we deduce from Corollary 13.6 that $\mathscr{I}(X, Y)$ is finite if X and Y are.

Before stating and proving the next theorem, we note that:

13.8 *If $X = \{x\}$ and Y is a finite set having $m(\in N)$ elements, then*

$$c(\mathscr{I}(X, Y)) = m = \frac{m!}{(m-1)!}.$$

Proof.—Since X has only one element, every mapping of X into Y is an injection; whence $\mathscr{I}(X, Y) = \mathscr{F}(X, Y)$. If $Y = \{y_1 \ldots, y_m\}$ (with $y_i \neq y_j$ if $i \neq j$), then $f \in \mathscr{F}(X, Y)$ if and only if there is $1 \leq j \leq m$ such that

$$f = (\{x\}, Y, \{(x, y_j)\}).$$

We deduce that $c(\mathscr{F}(X, Y)) = m$; hence 13.8 is proved.

13.9 Theorem.—*Let X and Y be finite sets such that $1 \leq c(X) \leq c(Y)$. Then*

13.10 $$c(\mathscr{I}(X, Y)) = \frac{c(Y)!}{(c(Y) - c(X))!}.$$

Proof.—We shall prove this theorem by mathematical induction. We reason as follows:

Let S be the set of all $m \in N$ such that whenever X is a finite set having m elements and Y is a finite set having $n \geq m$ elements, we have 13.10.

By 13.8, $1 \in S$. Suppose now that $m \in S$, and let X be a finite set having $m + 1$ elements and Y be a finite set having $n \geq m + 1$ elements. Let $x_0 \in X$ and $X' = X - \{x_0\}$; then X' has m elements.

Clearly, $f \mapsto f \mid X'$ is a mapping of $\mathscr{I}(X, Y)$ into $\mathscr{I}(X', Y)$; denote this mapping by φ. If $g \in \mathscr{I}(X', Y)$, then $g(X') \subset Y$ has m elements. Hence, there is $b \in Y - g(X')$. If we define $f : X \to Y$ by

$$
\textbf{13.11} \qquad \begin{aligned} f(x) &= b \quad \text{if} \quad x = x_0 \\ &= g(x) \quad \text{if} \quad x \in X', \end{aligned}
$$

then $f \in \mathscr{I}(X, Y)$ and $f \mid X' = g$. Therefore φ is a surjection.

Now let $g \in \mathscr{I}(X', Y)$. If $f \in \varphi^{-1}(g)$, then clearly $f \mid X' = g$ and $f(x_0) \in Y - g(X')$. Conversely, if $b \in Y - g(X')$ and if f is defined by 13.11, then $f \in \mathscr{I}(X, Y)$ and $\varphi(f) = g$. This shows that

$$
c(\varphi^{-1}(g)) = c(Y - g(X')).
$$

By Theorem 13.7, we have (note that $c(Y - g(X')) = c(Y) - c(g(X'))$ and $c(g(X') = c(X'))$)

$$
\begin{aligned}
c(\mathscr{I}(X, Y)) &= c(\mathscr{I}(X', Y)) c(Y - g(X')) \\
&= \frac{c(Y)!}{(c(Y) - c(X'))!} c(Y - g(X')) \\
&= \frac{c(Y)!}{(c(Y) - c(X') - 1)!} = \frac{c(Y)!}{(c(Y) - c(X))!}.
\end{aligned}
$$

We conclude that $m + 1 \in S$, hence, by mathematical induction, $S = N$. Thus, the proof is complete.

If X is a set, we denote by Σ_X the set of all permutations of X. Recall that a permutation of X is a bijection of X into X. If $X = \{1, 2, \ldots, n\}$, we shall sometimes write Σ_n instead of Σ_X.

13.12 Corollary.—*If X is a set having $m \in N$ elements, then*

$$
c(\Sigma_X) = n!.
$$

Proof.—If X is a finite set, then any injection $f : X \to X$ is a bijection; hence $\Sigma_X = \mathscr{I}(X, X)$. By Theorem 13.9, we have

$$
c(\Sigma_X) = c(\mathscr{I}(X, X)) = \frac{c(X)!}{(c(X) - c(X))!} = \frac{c(X)!}{0!} = c(X)!.
$$

For every set Y and integer $p \geq 0$, we denote by $\mathscr{P}_p(Y)$ the set of all parts of Y having p elements. Note that if $p = 0$, then $\mathscr{P}_p(Y) = \{\varnothing\}$; hence in this case $\mathscr{P}_p(Y)$ is a finite set having one and only one element.

13.13 Theorem.—*For every finite set Y having n elements and integer p satisfying $0 \leq p \leq n$, we have*

13.14
$$c(\mathscr{P}_p(Y)) = \frac{n!}{p!(n-p)!}.$$

Proof.—If $p = 0$, then $c(\mathscr{P}_p) = 1$ and

$$\frac{n!}{0!(n-0)!} = \frac{n!}{n!} = 1;$$

hence 13.14 holds in this case. Suppose now $p \geq 1$ and let $X = \{1, 2, \ldots, p\}$. For every $f \in \mathscr{I}(X, Y)$, the set $f(X) \in \mathscr{P}_p(Y)$. If we write

$$\varphi(f) = f(X) \quad \text{for} \quad f \in \mathscr{I}(X, Y),$$

we obtain a mapping $f \mapsto f(X)$ of $\mathscr{I}(X, Y)$ into $\mathscr{P}_p(Y)$. This mapping is a *surjection*. In fact, let $A = \{y_1, \ldots, y_p\} \in \mathscr{P}_p(Y)$ (we have $y_i \neq y_j$ if $i \neq j$); if we define $f : X \to Y$ by

$$f(j) = y_j \quad \text{for} \quad 1 \leq j \leq p,$$

then $f \in \mathscr{I}(X, Y)$ and $f(X) = A$.

Now let $A \in \mathscr{P}_p(Y)$ and consider the set $\varphi^{-1}(A)$. Note that $f \in \varphi^{-1}(A)$ if and only if $f \in \mathscr{I}(X, Y)$ and $f(X) = A$. Obviously, then, the number of elements of $\varphi^{-1}(A)$ is the same as the number of elements of the set $\mathscr{I}(X, A)$. By Theorem 13.9, this set has $p!$ elements. By Theorem 13.7, we deduce

$$c(\mathscr{I}(X, Y)) = p!c(\mathscr{P}_p(Y));$$

therefore

$$c(\mathscr{P}_p(Y)) = \frac{1}{p!} \frac{n!}{(n-p)!} = \frac{n!}{p!(n-p)!}.$$

Hence the theorem is proved.

Remark.—If $Y = \{\varnothing\}$ and $p = 0$, 13.14 still holds. In fact, in this case, $c(\mathscr{P}_p(Y)) = 1$ and

$$\frac{0!}{0!(0-0)!} = \frac{1}{1 \cdot 1} = 1.$$

For every two integers, p and n, such that $0 \leq p \leq n$, we write

$$\binom{n}{p} = \frac{n!}{p!(n-p)!} \quad \text{or} \quad \mathbf{C}_n^p = \frac{n!}{p!(n-p)!}.$$

These numbers are sometimes called *binomial coefficients*.

13.15 Corollary.—*For every two integers, n and p, satisfying $0 \leq p \leq n$, we have*

$$2^n = \binom{n}{0} + \binom{n}{1} + \ldots + \binom{n}{n}.$$

Proof.—Let $X = \{1, 2, \ldots, n\}$. Clearly

$$\mathscr{P}(X) = \mathscr{P}_0(X) \cup \mathscr{P}_1(X) \cup \ldots \cup \mathscr{P}_n(X).$$

Since $\mathscr{P}_i(X) \cap \mathscr{P}_j(X) = \varnothing$ if $i \neq j$, the result in Corollary 13.15 follows immediately from Theorems 13.4 and 13.14.

Let X be a set. A mapping $\sigma : X \to X$ is a *transposition* of X if and only if there exists a set $A = \{a, b\} \subset X$ consisting of two elements $a \neq b$ such that:

(i) $\sigma(a) = b$ and $\sigma(b) = a$;
(ii) $\sigma(x) = x$ if $x \in X - \{a, b\}$.

Note that the existence of a transposition of X implies that X contains at least two elements. The set of all *transpositions* of X will be denoted by J_X.

Note that if σ is the transposition defined by (i) and (ii), then

$$A = X - \{x \mid \sigma(x) = x\}.$$

Let σ be a transposition of X. Then

13.16 σ *is a permutation.*

(It follows from [i] and [ii] that if σ is a transposition of X, then σ is an *injection* and a *surjection*. Hence σ is a *bijection*.)

13.17 $\sigma^{-1} = \sigma.$

In fact, we have

$$\sigma(\sigma(x)) = x \quad \text{if} \quad x \in X - \{a, b\}$$

and

$$\sigma(\sigma(a)) = \sigma(b) = a, \qquad \sigma(\sigma(b)) = \sigma(a) = b.$$

Hence

$$\sigma(\sigma(x)) = x \quad \text{for all} \quad x \in X.$$

By Theorem 13.2 and the definition of σ^{-1}, it follows that $\sigma^{-1} = \sigma$.

13.18 Theorem.—*Let X be a set and for each $\sigma \in J_X$ let*

$$\varphi(\sigma) = X - \{x \mid \sigma(x) = x\}.$$

Then $\sigma \mapsto \varphi(\sigma)$ is a bijection of J_X into $\mathscr{P}_2(X)$.

▼ *Proof.*—Clearly, φ is a mapping of J_X into $\mathscr{P}_2(X)$. Now let σ_1 and σ_2 be two permutations and let

$$A_1 = X - \{x \mid \sigma_1(x) = x\} \quad \text{and} \quad A_2 = X - \{x \mid \sigma_2(x) = x\}.$$

If $A_1 = A_2 = \{\alpha, \beta\}$, then $\sigma_1(\alpha) = \beta$, $\sigma_1(\beta) = \alpha$, $\sigma_2(\alpha) = \beta$, $\sigma_2(\beta) = \alpha$; we deduce $\sigma_1 = \sigma_2$. Hence φ is an *injection*. From the definition of a transposition, it follows immediately that φ is a surjection. Hence φ is a bijection. ▲

13.19 Corollary.—*If X is a finite set having $n \geq 2$ elements, then*

$$c(J_X) = \tfrac{1}{2}n(n - 1).$$

Proof.—By Theorems 13.18 and 13.14, we have

$$c(J_X) = c(\mathscr{P}_2(X)) = \frac{n!}{2!(n - 2)!} = \tfrac{1}{2}n(n - 1).$$

Example.—Determine the transpositions of the set $X = \{1, 2, 3\}$. Let

$$\sigma_1 = \begin{pmatrix} 1 & 2 & 3 \\ 2 & 1 & 3 \end{pmatrix}, \qquad \sigma_2 = \begin{pmatrix} 1 & 2 & 3 \\ 3 & 2 & 1 \end{pmatrix}, \qquad \sigma_3 = \begin{pmatrix} 1 & 2 & 3 \\ 1 & 3 & 2 \end{pmatrix}.$$

Clearly, σ_1, σ_2, and σ_3 are (distinct) transpositions of $\{1, 2, 3\}$. By Corollary 13.19

$$c(J_X) = \tfrac{1}{2}3(3 - 1) = 3.$$

Hence $J_X = \{\sigma_1, \sigma_2, \sigma_3\}$.

13.20 Theorem.—*Let $p \in N$ and $q \in N$ and let $\sum_{p,q}$ be the set of all permutations σ of $X = \{1, \ldots, p + q\}$ satisfying*

$$\sigma(1) < \ldots < \sigma(p) \quad \text{and} \quad \sigma(p + 1) < \ldots < \sigma(p + q).$$

Then

$$c\left(\textstyle\sum_{p,q}\right) = \frac{(p + q)!}{p!\,q!}.$$

Proof.—Let $A \in \mathscr{P}_p(X)$ and let

$$A = \{i_1, \ldots, i_p\} \quad \text{with} \quad i_1 < i_2 < \ldots < i_p$$

and
$$X - A = \{i_{p+1}, \ldots, i_{p+q}\} \quad \text{with} \quad i_{p+1} < \ldots < i_{p+q}.$$

Now let $\sigma_A \in \sum_X$ be the permutation

$$\begin{pmatrix} 1 \ldots p & p+1 \ldots p+q \\ i_1 \ldots i_p & i_{p+1} \ldots i_{p+q} \end{pmatrix}.$$

Clearly, $\sigma_A \in \sum_{p,q}$. Note also that $\sigma_A(\{1, \ldots, p\}) = A$. If $A \in \mathscr{P}_p(X)$ and $B \in \mathscr{P}_p(X)$ then

$$\sigma_A(\{1, \ldots, p\}) = A \quad \text{and} \quad \sigma_B(\{1, \ldots, p\}) = B.$$

Hence, $\sigma_A \neq \sigma_B$ if $A \neq B$. Hence $A \mapsto \sigma_A$ is an injection of $\mathscr{P}_p(X)$ into $\sum_{p,q}$.

Now let $\sigma \in \sum_{p,q}$ and let

$$A = \{\sigma(1), \ldots, \sigma(p)\}.$$

Then $A \in \mathscr{P}_p(X)$. Since

$$\sigma(1) < \ldots < \sigma(p) \quad \text{and} \quad \sigma(p+1) < \ldots < \sigma(p+q),$$

we deduce that $\sigma_A = \sigma$. Hence, $A \mapsto \sigma_A$ is a surjection and therefore a *bijection*. Hence, there is a bijection of $\mathscr{P}_p(X)$ onto $\sum_{p,q}$, and hence

$$c\left(\sum_{p,q}\right) = \frac{(p+q)!}{p!((p+q)-p)!} = \frac{(p+q)!}{p!q!}.$$

Exercises for Chapter 13

1. Compute:

   (a) $\binom{5}{2}$;

   (b) $c(J_X)$ where $X = \{a, 3, \varnothing\}$;

   (c) $\binom{7}{3}$;

   (d) $c(\mathscr{I}(X, Y))$ where X is as in 1(b) and $Y = A_4$ (see Exercise 3, Chapter 1);

   (e) $c(\mathscr{P}(A_4))$;

   (f) $c(A_3)$.

2. Show that $\binom{n}{p} = \binom{n}{n-p}$ for $n \in N$, $1 \leq p \leq n$.

3. Show that for $n \in N$, $1 \leq p \leq n$

$$\binom{n}{p-1} + \binom{n}{p} = \binom{n+1}{p}.$$

APPENDICES

Real Numbers

The purpose of this appendix is to gather some of the terminology we use and to state some of the properties of real numbers.

The set of real numbers is denoted by R. The set R is endowed with two operations, an *addition* and a *multiplication*. The addition associates to every pair (x, y) of real numbers the *sum* $x + y$; the multiplication associates to every pair (x, y) of real numbers the *product* xy. These operations have the following basic properties (x, y, z, belong to R):

(1) $x + y = y + x$ (commutativity of addition);
(2) $x + (y + z) = (x + y) + z$ (associativity of addition);
(3) $x + 0 = x$ (existence of the zero element);
(4) $x + (-x) = 0$ (existence of the inverse element for addition);
(5) $xy = yx$ (commutativity of multiplication);
(6) $x(yz) = (xy)z$ (associativity of multiplication);
(7) $1x = x$ (existence of the identity element);
(8) if $x \neq 0$, then $xx^{-1} = 1$ (existence of the inverse element for multiplication);
(9) $x(y + z) = xy + xz$ (distributivity of multiplication with respect to addition).

Let x and y be real numbers. We may write $x \cdot y$ instead of xy. We also write $x - y = x + (-y)$ and $-x - y = (-x) + (-y)$. If $y \neq 0$, we write $1/y = y^{-1}$.

The set of the *positive real numbers* will be denoted R_+. For $x \in R$, the notation $x \geq 0$ (read, "x greater than or equal to 0") is equivalent to $x \in R_+$. The set R_+ has the properties:

(10) $0 \in R_+$;

(11) if $x \in R_+$ and $y \in R_+$ then $x + y \in R_+$ and $xy \in R_+$;

(12) If $x \in R$ then $x \in R_+$ and $-x \in R_+$ if and only if $x = 0$;

(13) if $x \in R$ then either $x \in R_+$ or $-x \in R_+$.

Now let x and y be in R. We write $x \geq y$ (read, "x greater than or equal to y") or $y \leq x$ (read, "y less than or equal to x") if and only if $x - y \in R_+$. A real number, x, is *negative* if $-x \in R_+$. It follows from (13) that every real number is either positive or negative. From (12), it follows that 0 is both positive and negative and is the only real number that has both these properties. We write $x > y$ (read, "x greater than y") or $y < x$ (read, "y less than x") if $x \geq y$ and $x \neq y$. A real number, x, is *strictly positive* if $x > 0$; a real number is *strictly negative* if $x < 0$.

The properties (1)–(9) of R express that R is a "field." The properties (1)–(13) express that R is an "ordered field."

The set of all integers $\ldots -2, -1, 0, 1, 2, \ldots$ is denoted Z. The set of strictly positive integers $1, 2, 3, \ldots$ is denoted N; the elements of N are called natural numbers.*

A real number is *rational* if it can be written in the form n/m with $n \in Z$, $m \in N$. The set of all rational numbers is denoted Q. The set of all positive rational numbers is denoted Q_+. The set of all positive integers is denoted Z_+. Notice that $Z_+ = N \cup \{0\}$; $Q_+ = R_+ \cap Q$, and $Z_+ = R_+ \cap Z = Q_+ \cap Z$.

The properties (1)–(13) together with the property (14) below "characterize" the real numbers.

(14) *Every non-void set $A \subset R$, bounded above, has a supremum.*

We wish to point out that we did not prove here that there exists a set R having the properties (1)–(14). Such a theorem can be proved, but the various proofs that have been devised are long and involved. The reader interested in a proof may consult, for instance, [4] or [7]. A new method of proof is developed in [9].

From (1)–(14) we can deduce several other properties, among which we shall give the following without proof.†

I.1 *Every non-void set, $A \subset R$, bounded below has an infimum.*

---

* N can be shown to be the intersection of all subsets T of R such that $1 \in T$ and $x \in T \Rightarrow x + 1 \in T$.

† With the exception of I.6.

We notice that A is bounded below if and only if $-A$ is bounded above, and that inf $A = -\sup(-A)$.

I.2   The Property of Archimedes.—*Let $x \in R$, $x > 0$. For any $y \in R$, there exists $n \in N$ such $y < nx$.*

▼   The property of Archimedes is essential in proving that the sequence $(1/n)_{n \in N}$ converges to zero.   ▲

I.3   *For every $a \in R$, $b \in R$, $a < b$ there exists $r \in Q$ satisfying $a < r < b$.*

For any $a \in R$, $b \in R$, $a < b$ we shall write

$$[a, b] = \{x \mid a \le x \le b\}$$

We call such a set a *closed (bounded) interval.*

I.4   (*The nested interval property*) *Let $(I_n)_{n \in N}$ be a sequence of closed intervals such that*

$$I_1 \supset I_2 \supset I_3 \supset \ldots \supset I_n \supset \ldots .$$

Then there exists $\alpha \in R$ such that $\alpha \in I_n$ for all $n \in N$.

With the notations of Chapter 8, the conclusion of I.4 can be expressed by writing $\bigcap_{n \in N} I_n \ne \varnothing$.

I.5   *For any $p \in N$ and $y \in R_+$ there exists a unique $x \in R_+$ such that*
$$x^p = y.$$

We usually write $x = y^{1/p} = \sqrt[p]{y}$ and call x the *root of order p* of y. For $p = 2$, we write $\sqrt{y}$ instead of $\sqrt[2]{y}$.

I.6   The Theorem of Pythagoras.—*There is no $r \in Q$ satisfying $r^2 = 2$.*

The proof makes use of the following result: If p is an integer such that p^2 is even,* then p is even. In fact, suppose p^2 is even. If p is not even, then p is odd; that is, $p = 2n + 1$ for some $n \in Z$. Then

$$p^2 = (2n + 1)^2 = 4n^2 + 4n + 1 = 2(2n^2 + 2n) + 1;$$

that is, p^2 is odd. But this leads to a contradiction. Since the hypothesis that p is not even leads to a contradiction, we conclude that the integer p is even.

---

* The integer p is even if $p = 2m$ for some $m \in Z$. The integer p is odd if $p = 2m + 1$ for some $m \in Z$.

Proof of I.6.—Suppose there were an $r \in Q$ satisfying $r^2 = 2$. Let $r = p/q$ where p and $q \neq 0$ are integers. We may and shall suppose that p and q *are not both even.* Since $r^2 = 2$, we have $p^2/q^2 = 2$; that is, $p^2 = 2q^2$. Hence, p^2 is even, and hence (by the previous remark) p is even. Hence, $p = 2m$ for some $m \in Z$. Then $4m^2 = p^2 = 2q^2$, whence $2m^2 = q^2$. Thus q^2 is even, which contradicts our assumption. Since the assumption that there is $r \in Q$ satisfying $r^2 = 2$ leads to a contradiction, it follows that there is no such $r \in Q$.

From I.5 and I.6, we deduce that $Q \neq R$. The real numbers in $R - Q$ are called *irrational numbers.*

The Signature of a Permutation

Let S be a set. We shall call *law of composition* on S (or binary operation on S) any mapping $(x, y) \mapsto \varphi(x, y)$ of $S \times S$ into S; the element $\varphi(x, y)$ will be called the composition of x and y (for the considered law of composition).

Example 1.—Let R be the set of real numbers. Then $(x, y) \mapsto x + y$ and $(x, y) \mapsto xy$ are laws of composition on R.

Example 2.—Let X be a set. Then $(A, B) \mapsto A \cup B$ and $(A, B) \mapsto A \cap B$ are laws of composition on $\mathscr{P}(X)$.

Example 3.—Let A be a set and $S = \Sigma_A$ ($=$ the set of all permutations of A). Then $(f, g) \mapsto f \circ g$ is a law of composition on S.

Example 4.—Let $R^* = \{x \mid x \in R, \ x \neq 0\}$. Then $(x, y) \mapsto x/y$ is a law of composition on R^*.

We shall now introduce the following:

II.1 Definition.—*A set G endowed with a law of composition $(x, y) \mapsto x \perp y$ is a group if:*
    (i) *$x \perp (y \perp z) = (x \perp y) \perp z$ for all $x \in G, y \in G, z \in G$;*
    (ii) *There exists $e \in G$ such that $x \perp e = e \perp x = x$ for all $x \in G$;*
    (iii) *For every $x \in G$, there is $x' \in G$ such that $x \perp x' = x' \perp x = e$.*

Remarks.*—(1) There exists only one element e in G satisfying II.1(ii). In fact, suppose that e' and e'' belong to G and that

$$x \perp e' = e' \perp x = x \quad \text{and} \quad x \perp e'' = e'' \perp x = x$$

for all $x \in G$. Then

$$e' = e' \perp e'' = e'' \Rightarrow e' = e''.$$

(2) Given $x \in G$, there is only one $x' \in$ satisfying II.1(iii). In fact,

$$x' = x' \perp e = x' \perp (x \perp x'') = (x' \perp x) \perp x'' = e \perp x'' = x'';$$

whence $x' = x''$.

The mapping $(x, y) \mapsto x \perp y$ is called the law of composition of the group. The property II.1(i) asserts that $(x, y) \mapsto x \perp y$ is *associative*. The property (ii) asserts that it has a *neutral element*, and the property (iii) asserts that every $x \in G$ is *invertible*. The element, x', corresponding to x is called the inverse of x. Hence, a group is a set endowed with a law of composition that is associative, that has a neutral element, and that is such that every $x \in G$ is invertible. Notice that if G is a group, then $G \neq \varnothing$, since G contains a neutral element.

The law of composition of a group is often written $(x, y) \mapsto xy$. In this case, we say that we use the *multiplicative* notation. The neutral element is then called the *unit element* of G and is denoted 1. When we use the multiplicative notation, the inverse of an element $x \in G$ is denoted x^{-1}.

The group G is *commutative* if $x \perp y = y \perp x$ for all $x \in G, y \in G$. The law of composition of a commutative group is often written $(x, y) \mapsto x + y$. In this case, we say that we use the *additive* notation. The neutral element is then called the *zero element* of G and is denoted 0. When we use the additive notation, the inverse of an element $x \in G$ is denoted $-x$. The use of the additive notation implies that the group is commutative.

Example 5.—The set S in Example 3 is a group. The unit element of S is j_A (recall that $j_A(x) = x$ for all $x \in A$). The inverse of $f \in \Sigma_A$ is the function f^{-1} (see Chapter 7).

Exercise.—Show that the group S in the above example is not commutative if A contains at least *three* distinct elements.

---

* In these remarks we assume that G is a group.

Example 6.—The set R^* endowed with the law of composition $(x, y) \mapsto xy$ is a (commutative) group. The unit element is the number 1. The inverse of $x \in R^*$ is the number $1/x$.

The groups in Examples 5 and 6 will be used in the following sections.

Let G and G' be two groups.* A mapping $f : G \to G'$ is a *representation* of G into G' if

II.2 $$f(xy) = f(x)f(y)$$

for all $x \in G$, $y \in G$.

Remarks.—Let $f : G \to G'$ be a representation. Let e be the unit element of G and e' the unit element of G'. Then:

II.3 $$f(e) = e';$$

II.4 $$f(x^{-1}) = f(x)^{-1} \quad \text{for all} \quad x \in G.$$

Proof of II.3.—We have

$$f(e) = e'f(e) = (f(e)^{-1}f(e))f(e)$$
$$= f(e)^{-1}(f(e)f(e)) = f(e)^{-1}f(ee)$$
$$= f(e)^{-1}f(e) = e'.$$

Proof of II.4.—Let $x \in G$. We have

$$f(x^{-1})f(x) = f(x^{-1}x) = f(e) = e'.$$

II.5   If $x_1, \ldots, x_n$ belong to G, then

$$f(x_1, \ldots, x_n) = f(x_1) \ldots f(x_n).$$

This can be proved by using mathematical induction and II.2, the defining property of a representation.

For each $n \in N$, let Σ_n be the group of permutations of the set $X = \{1, 2, \ldots, n\}$. The identity mapping of X onto X is denoted j_X; when $X = \{1, 2, \ldots, n\}$ we shall usually write j_n instead of j_X. We have (see Corollary 13.12)

$$c(\Sigma_n) = n!.$$

---

* We denote multiplicatively the laws of composition of the groups we consider here.

A mapping $\sigma : X \to X$ is a *transposition* of X if and only if there exists a set $\{a, b\} \subset X$, with $a \neq b$, such that:

(i) $\sigma(a) = b$ and $\sigma(b) = a$;
(ii) $\sigma(x) = x$ for all $x \in X - \{a, b\}$.

A transposition σ is a permutation and $\sigma^{-1} = \sigma$ (see 13.16 and 13.17).

The *signature of a permutation* $\sigma \in \Sigma_n$ is introduced in Theorem II.10. For its proof, we shall need two propositions, which we shall establish first.

II.6   Proposition.—*Let $n \in N$, $n \geq 2$, and let $\sigma \in \Sigma_n$. Then there exist transpositions $\tau_1, \tau_2, \ldots, \tau_p$ belonging to Σ_n such that*

II.7 $$\sigma = \tau_1 \circ \tau_2 \circ \ldots \circ \tau_p.$$

▼   *Proof.*—Notice first that if $\sigma \in \Sigma_n$ satisfies $\sigma(n) = n$, then there are transpositions $\alpha_1, \ldots, \alpha_q$ belonging to Σ_n such that if $u = \alpha_1 \circ \ldots \circ \alpha_q \circ \sigma$, then $u(n) = n$. In fact, for every $1 \leq j < n$, let s_j be the transposition* satisfying

II.8 $s_j(j) = j + 1.$

If $\sigma \in \Sigma_n$ and $\sigma(n) \neq n$, then $\sigma(n) = j < n$. Hence

$$s_{n-1} \circ \ldots \circ s_j \circ \sigma(n) = s_{n-1} \circ \ldots \circ s_j(j) = n$$

and thus $u(n) = n$ if $u = s_{n-1} \circ \ldots \circ s_j \circ \sigma$. If, moreover, $u = \gamma_1 \circ \ldots \circ \gamma_r$ when $\gamma_1, \ldots, \gamma_r$ are transpositions belonging to Σ_n, then

$$\gamma_1 \circ \ldots \circ \gamma_r = s_{n-1} \circ \ldots \circ s_j \circ \sigma$$

and therefore

$$\sigma = (s_{n-1} \circ \ldots \circ s_j)^{-1} \circ (\gamma_1 \circ \ldots \circ \gamma_r)$$

$$= s_1^{-1} \circ \ldots \circ s_{n-1}^{-1} \circ \gamma_1 \circ \ldots \circ \gamma_r.$$

We conclude that *to prove that an arbitrary permutation belonging to Σ_n can be written in the form II.7 it is enough to show that any $\sigma \in \Sigma_n$ satisfying $\sigma(n) = n$ can be written in the form II.7.*

Now let A be the set of all $p \in N$ such that if $\sigma \in \Sigma_{1+p}$ then σ can be represented in the form II.7.

---

* Notice that there is one and only one transposition, $s_j \in \Sigma_n$, satisfying II.8. We have $s_j(j + 1) = j$ and $s_j(m) = m$ for $m \neq j$, $m \neq j + 1$.

If $p = 1$, then $\Sigma_{1+p} = \Sigma_2$ consists of the permutations

$$j_2 = \begin{pmatrix} 1 & 2 \\ 1 & 2 \end{pmatrix} \quad \text{and} \quad \tau = \begin{pmatrix} 1 & 2 \\ 2 & 1 \end{pmatrix}.$$

Clearly, τ is a transposition, $j_2 = \tau \circ \tau$ and $\tau = \tau \circ \tau \circ \tau^{-1}$. Hence, $1 \in A$.

Suppose now that $p \in A$ and let $\sigma \in \Sigma_{1+(p+1)}$ be such that $\sigma(1 + (p + 1)) = 1 + (p + 1)$. Let $\sigma' \in \Sigma_{1+p}$ be the permutation

$$\begin{pmatrix} 1 & 2 & \dots & 1+p \\ \sigma(1) & \sigma(2) & \dots & \sigma(1+p) \end{pmatrix}.$$

Since $p \in A$, there are transpositions $\delta_1', \dots, \delta_r'$ belonging to Σ_{1+p} such that

$$\sigma' = \delta_1' \circ \dots \circ \delta_r'.$$

For each $1 \le j \le r$, now let

$$\delta_j = \begin{pmatrix} 1 & 2 & \dots & 1+p & 1+(p+1) \\ \delta_j'(1) & \delta_j'(2) & \dots & \delta_j'(1+p) & 1+(p+1) \end{pmatrix};$$

then $\delta_j \in \Sigma_{1+(p+1)}$ and is a transposition. Clearly, also,

$$\sigma = \delta_1 \circ \dots \circ \delta_r.$$

Since $\sigma \in \Sigma_{1+(p+1)}$ was arbitrary, except for the hypothesis

$$\sigma(1 + (p + 1)) = 1 + (p + 1),$$

we deduce that $1 + (p + 1) \in A$.

By mathematical induction, $A = N$. Thus, Proposition II.6 is proved.   ▲

Let $n \ge 2$ and $I = \{(i,j) \mid 1 \le i < j \le n\}$. For $\sigma \in \Sigma_n$, let

$$A(\sigma) = \{(i,j) \mid (i,j) \in I, \sigma(i) < \sigma(j)\}$$

and

$$B(\sigma) = \{(i,j) \mid (i,j) \in I, \sigma(j) < \sigma(i)\}.$$

Clearly

$$A(\sigma) \cap B(\sigma) = \varnothing \quad \text{and} \quad A(\sigma) \cup B(\sigma) = I.$$

For $\sigma \in \Sigma_n$, let $T_\sigma : I \to I$ be defined by

$$T_\sigma(i,j) = \begin{cases} (\sigma(i), \sigma(j)) & \text{if} \quad (i,j) \in A(\sigma) \\ (\sigma(j), \sigma(i)) & \text{if} \quad (i,j) \in B(\sigma). \end{cases}$$

We leave it to the reader to verify that T_σ is a *bijection*.

For any $f \in \Sigma_n$, we define

$$F(f) = \prod_{(i,j) \in I} (f(j) - f(i))$$

and

$$\mathscr{E}(\sigma) = (-1)^{c(B(\sigma))}$$

for every $\sigma \in \Sigma_n$. Then:

II.9   Proposition.—*For every σ and f in Σ_n, we have*

$$F(f \circ \sigma) = \mathscr{E}(\sigma) F(f).$$

▼   *Proof.*—Let $\tilde{f}$ be the mapping $(i,j) \mapsto f(j) - f(i)$ of $X \times X$ into $\mathbf{R}$. Then, if σ and f are in Σ_n, we have

$$F(f \circ \sigma) = \prod_{(i,j) \in I} (f(\sigma(j)) - f(\sigma(i))) = \prod_{(i,j) \in I} \tilde{f}(\sigma(i), \sigma(j))$$

$$= \prod_{(i,j) \in A(\sigma)} \tilde{f}(\sigma(i), \sigma(j)) \prod_{(i,j) \in B(\sigma)} \tilde{f}(\sigma(i), \sigma(j))$$

$$= \prod_{(i,j) \in A(\sigma)} \tilde{f}(\sigma(i), \sigma(j))((-1)^{c(B(\sigma))}) \prod_{(i,j) \in B(\sigma)} \tilde{f}(\sigma(j), \sigma(i))$$

$$= \mathscr{E}(\sigma) \prod_{(i,j) \in A(\sigma)} \tilde{f}(T_\sigma(i,j)) \prod_{(i,j) \in B(\sigma)} \tilde{f}(T_\sigma(i,j))$$

$$= \mathscr{E}(\sigma) \prod_{(i,j) \in I} \tilde{f}(T_\sigma(i,j)) = \mathscr{E}(\sigma) \prod_{(s,t) \in T_\sigma(I)} \tilde{f}(s, t)$$

$$= \mathscr{E}(\sigma) \prod_{(s,t) \in I} \tilde{f}(s, t) = \mathscr{E}(\sigma) F(f).$$

Since σ and f were arbitrary, the proposition is proved.   ▲

We now give the main result of this Appendix.

II.10   Theorem.—*Let $n \in \mathbf{N}$. Then there exists one and only one representation $\mathscr{E} : \Sigma_n \to \mathbf{R}^*$ such that*

II.11 $\mathscr{E}(\sigma) = -1$ *whenever σ is a transposition.*

Moreover, the image of Σ_n by $\mathscr{E}$ is $\{-1, +1\}$.

For each $\sigma \in \Sigma_n$, the number $\mathscr{E}(\sigma)$ is called the signature of σ. By II.11, the signature of a transposition is -1. Since $\mathscr{E}$ is a representation, the signature of j_n is 1.

A permutation σ is said to be *even* if $\mathscr{E}(\sigma) = 1$; note that if σ and ρ are even, then $\sigma \circ \rho$ is even. A permutation σ is said to be *odd* if $\mathscr{E}(\sigma) = -1$; note that if σ and ρ are odd, then $\sigma \circ \rho$ is even.

If ρ is a permutation, then

$$\mathscr{E}(\rho)(\rho^{-1}) = \mathscr{E}(\rho \circ \rho^{-1}) = \mathscr{E}(j_n) = 1.$$

It follows that ρ is even if and only if ρ^{-1} is even; ρ is odd if and only if ρ^{-1} is odd.

▼   *Proof of Theorem* II.10.—Define $\mathscr{E}: \Sigma_n \to \boldsymbol{R}^*$ by

$$\mathscr{E}(\sigma) = (-1)^{c(B(\sigma))}$$

for $\sigma \in \Sigma_n$. Let f be the identity mapping j_n of X onto X and let σ and ρ be elements of Σ_n. By Proposition II.9,

$$\mathscr{E}(\sigma \circ \rho)F(f) = F(f \circ (\sigma \circ \rho))$$
$$= F((f \circ \sigma) \circ \rho) = \mathscr{E}(\rho)F(f \circ \sigma) = \mathscr{E}(\sigma)\mathscr{E}(\rho)F(f).$$

Since, clearly, $F(f) \neq 0$, we deduce $\mathscr{E}(\sigma \circ \rho) = \mathscr{E}(\sigma)\mathscr{E}(\rho)$. Since σ and ρ in Σ_n were arbitrary, it follows that $\mathscr{E}$ *is a representation of Σ_n into $\boldsymbol{R}^*$.*

We shall show now that II.11 is also satisfied. To do this, it is enough to show that $c(B(\sigma))$ is *odd* if σ is a transposition. In fact, if σ is a transposition, then there is $(i, j) \in I$ such that $\sigma(i) = j$, $\sigma(j) = i$, and $\sigma(h) = h$ if $h \notin \{i, j\}$. Let

$$Y = \{(i, h) \mid i < h < j\} \cup \{(h, j) \mid i < h < j\} \cup \{(i, j)\}.$$

It is clear that if $(u, v) \in Y$, then $\sigma(u) > \sigma(v)$; hence $Y \subset B(\sigma)$. Now let

$$Z = \{(u, v) \in I \mid u < i\} \cup \{(u, v) \in I \mid j < v\} \cup \{(u, v) \mid i < u < v < i\}.$$

Then $Z \cap B(\sigma) = \varnothing$ and $Y \cup Z = I$. Since $Y \subset B(\sigma)$, we deduce $Y = B(\sigma)$. Now $\{(i, h) \mid i < h < j\}$ and $\{(h, j) \mid i < h < j\}$ have the same number of elements. Hence, $c(B(\sigma))$ is *odd*.

The uniqueness assertion in the theorem can be proved as follows: Let $\mathscr{E}'$ and $\mathscr{E}''$ be two representations of Σ_n into $\boldsymbol{R}^*$ having the property II.11. If $n = 1$, then Σ_n contains only the permutation j_n, and hence in this case $\mathscr{E}' = \mathscr{E}''$. Let $n \geq 2$ and let $\sigma \in \Sigma_n$. By II.6

$$\sigma = \tau_1 \circ \tau_2 \circ \ldots \circ \tau_p$$

and $\tau_1, \tau_2, \ldots, \tau_p$ are transpositions. By II.6,

$$\mathscr{E}'(\sigma) = \mathscr{E}'(\tau_1)\mathscr{E}'(\tau_2) \ldots \mathscr{E}'(\tau_p) = (-1)^p$$

and

$$\mathscr{E}''(\sigma) = \mathscr{E}''(\tau_1)\mathscr{E}''(\tau_2) \ldots \mathscr{E}''(\tau_p) = (-1)^p.$$

Hence, $\mathscr{E}'(\sigma) = \mathscr{E}''(\sigma)$. Since σ was arbitrary, *we conclude $\mathscr{E}' = \mathscr{E}''$.*

Since the image of Σ_n by $\mathscr{E}$ is obviously $\{-1, +1\}$, the theorem is completely proved ▲

The representation of a permutation as a product of transpositions (see Proposition II.6) is not unique. In fact, if $\sigma = \tau_1 \circ \tau_2 \circ \ldots \circ \tau_p$ is such a representation for σ, and if τ is a transposition, then τ^{-1} is also a transposition and:

$$\sigma = \tau_1 \circ \tau_2 \circ \ldots \circ \tau_p \circ \tau \circ \tau^{-1}.$$

However, we have the following corollary:

II.12   Corollary.—*Let $\sigma \in \Sigma_n$ be a permutation and let*

$$\sigma = \alpha_1 \circ \ldots \circ \alpha_n = \beta_1 \circ \ldots \circ \beta_m \qquad (n \in \mathbf{N}, m \in \mathbf{N})$$

where $\alpha_1, \ldots, \alpha_n, \beta_1, \ldots, \beta_m$ are transpositions. Then m is even if and only if n is even (whence m is odd if and only if n is odd).

Proof.—Since $\mathscr{E}$ is a representation, and since II.11 is satisfied, we have

$$\mathscr{E}(\sigma) = (-1)^n \quad \text{and} \quad \mathscr{E}(\sigma) = (-1)^m.$$

Hence, $(-1)^n = (-1)^m$, and hence m is even $\Leftrightarrow$ n is even.

Let G be a *commutative group*, the law of composition of which is written additively. If $x \in G$, then $(-1)x = -x$ and $1x = x$; recall that $(-1)((-1)x) = x$. If $\sigma \in \Sigma_n$, then $\mathscr{E}(\sigma) = -1$ or 1; hence it makes sense to write $\mathscr{E}(\sigma)x$ if $\sigma \in \Sigma_n$ and $x \in G$.

Now let S be a set. A mapping $f:S^n \to G$ (see the end of Chapter 3 for the notation S^n) is said to be *antisymmetric* if

$$f(x_{\sigma(1)}, \ldots, x_{\sigma(n)}) = \mathscr{E}(\sigma)f(x_1, \ldots, x_n)$$

for all $\sigma \in \Sigma_n$ and $(x_1, \ldots, x_n) \in S^n$.

We leave it to the reader to show that a mapping $f:S^n \to G$ is antisymmetric if and only if

$$f(x_{\sigma(1)}, \ldots, x_{\sigma(n)}) = -f(x_1, \ldots, x_n)$$

for all transpositions $\sigma \in \Sigma_n$ and $(x_1, \ldots, x_n) \in S^n$ (use Proposition II.6).

Example 7.—Let R be endowed with the composition law $(x, y) \mapsto x + y$; then R is a commutative group. For every $(x_1, x_2) \in R^2$ and $(y_1, y_2) \in R^2$, *define*

$$\begin{vmatrix} x_1 & x_2 \\ y_1 & y_2 \end{vmatrix} = x_1 y_2 - x_2 y_1.$$

Then the mapping

$$((x_1, x_2), (y_1, y_2)) \mapsto \begin{vmatrix} x_1 & x_2 \\ y_1 & y_2 \end{vmatrix}$$

of $R^2 \times R^2$ into R is antisymmetric.

Exercise.—Show that

$$\sum_{\sigma \in \Sigma_n} \mathscr{E}(\sigma) = 0 \quad (\text{if} \quad n \geq 2).$$

Supplementary Exercises

1. Let G be a set. We say that G is a *graph* if and only if every element of G is an ordered pair; i.e.,

G is a graph $\Leftrightarrow z \in G \Rightarrow z = (x, y)$ for some x and some y.

Is the set $\{(a, b), (1, 2), (1, 3)\}$ a graph?
Is the set $\{(a, b), 2\}$ a graph?
Is $\{((a, b), 2)\}$ a graph?
Is $((a, b), 2)$ a graph?

2. Let G be a graph. We denote by $pr_1(G)$ the set

$$\{x \mid \text{there exists } y \text{ such that } (x, y) \in G\}.$$

We denote by $pr_2(G)$ the set

$$\{y \mid \text{there exists } x \text{ such that } (x, y) \in G\}.$$

Note that $pr_1(G)$ is "the set of all first components of elements of G" and $pr_2(G)$ is "the set of all second components of elements of G." Let:

$$G_1 = \{(a, 1), (3, (3, 4))\};$$

$$G_2 = \{(1, 2), (1, 3), (1, 4)\};$$

$$G_3 = \{(a, (1, 2)), (a, (2, 3))\}.$$

Construct $pr_i(G_j)$ for $i = 1, 2$ and $j = 1, 2, 3$. Construct $pr_1(pr_2(G_3))$.

3. Let G be a graph, and X be a set. We denote by $G[X]$ the set

$$\{y \mid \text{there exists } x \in X \text{ such that } (x, y) \in G\}.$$

Let G_1, G_2, G_3 be as in Exercise 2 . Let $X_1 = \{1, 3\}$ and $X_2 = \{a\}$. Construct $G_i[X_j]$ for $i = 1, 2, 3$ and $j = 1, 2$.

4. Let G be a graph. Define $G^{-1} = \{(y, x) \mid (x, y) \in G\}$. Construct G_i^{-1}, $i = 1, 2, 3$ (see Exercise 2). Show that $(G^{-1})^{-1} = G$.

5. Let G be a graph and A be a set. Show that

$$G^{-1}[A] = \{s \mid \text{there exists } t \in A \text{ such that } (s, t) \in G\}.$$

6. Let G be a graph. Show that $G^{-1}[G[pr_1(G)]] = pr_1(G)$.

7. Let R and S be graphs and let A and B be sets. Show that

    (a) $R \subset S \Rightarrow R[A] \subset S[A]$;
    (b) $A \subset B \Rightarrow R[A] \subset R[B]$;
    (c) $R \subset S \Rightarrow R^{-1} \subset S^{-1}$;
    (d) $R[A \cap B] \subset R[A] \cap R[B]$;
    (e) $R[A \cup B] = R[A] \cup R[B]$;
    (f) $R^{-1}[R[A]] \supset A$ if $A \subset pr_1(R)$.

8. Let R and S be graphs. Define a graph $R \circ S$ to be the set

$$\{(x, z) \mid \text{there is } y \text{ such that } (x, y) \in S \text{ and } (y, z) \in R\}.$$

Construct $R \circ S$ and $S \circ R$ in each of the following cases:

    (a) $R = \{(0, 1), (1, 2)\}$, $S = \{(1, 2), (2, 3)\}$;
    (b) $R = \{(0, 1), (1, 2)\}$, $S = \{(0, 1), (1, 2)\}$;
    (c) $R = \{(0, 0), (1, 2)\}$, $S = \{(1, 2), (1, 0)\}$;
    (d) $R = \{(0, 1), (0, 2), (0, 3)\}$, $S = \{(1, 0), (2, 0), (3, 0)\}$;
    (e) Show that if R and S are any two graphs, then

$$R \circ S = \{(x, y) \mid S[\{x\}] \cap R^{-1}[\{y\}] \neq \varnothing \}.$$

9. Let G_1, G_2, and G_3 be as in Exercise 2 .

    (a) Construct $G_i \circ G_j$ for $i = 1, 2, 3$, and $j = 1, 2, 3$;
    (b) Construct $G_1^{-1} \circ G_3$, $G_1 \circ G_3^{-1}$, $G_3 \circ G_1^{-1}$, $G_3^{-1} \circ G_1$;
    (c) Construct $(G_2 \circ G_1)^{-1}$ and $G_1^{-1} \circ G_2^{-1}$;
    (d) Construct $G_3 \circ (G_1^{-1} \circ G_2)$ and $(G_3 \circ G_1^{-1}) \circ G_2$.

10. Let R, S, and T be graphs, and let A be a set. Show that:

    (a) $R \subset S \Rightarrow R \circ T \subset S \circ T$;
    (b) $R \subset S \Rightarrow T \circ R \subset T \circ S$;
    (c) $R \circ S[A] = R[S[A]]$.

11. Let R, S, and T be graphs. Show that:

    (a) $(R \circ S)^{-1} = S^{-1} \circ R^{-1}$;

    (b) $(R \circ S) \circ T = R \circ (S \circ T)$.

12. Let G be a graph. We say that G is a *functional graph* if and only if for every x, $G[\{x\}]$ has at most one element. Let G_1, G_2, and G_3 be as in Exercise 2 . Which of these are functional graphs?

13. For any set X, denote by $\Delta(X)$ the set $\{(t, t) \mid t \in X\}$ (see also Chapter 4). Let R be a graph. Show that:

$$R \text{ is a functional graph} \iff R \circ R^{-1} \subset \Delta(pr_2(R)).$$

14. Let R and S be functional graphs. Show that $R \circ S$ is a functional graph. Show also that $R \cap S$ is a functional graph.

15. Give an example of two graphs, R and S, such that R is not a functional graph and S is not a functional graph, but $R \circ S$ is a functional graph.

16. Give an example of two functional graphs, R and S, such that $R \cup S$ is not a functional graph.

17. Let R and S be two functional graphs. Find a condition [   ] involving R and S such that [   ] $\Rightarrow R \cup S$ is a functional graph. (There are several answers.)

18. A *correspondence* is any triple (A, B, S) that satisfies:

    (i) A, B and S are sets;

    (ii) $S \subset A \times B$.

Let (C, D, T) be given. Show that (C, D, T) is a correspondence if and only if:

    (a) C and D are sets;

    (b) T is a graph;

    (c) $pr_1(T) \subset C$ and $pr_2(T) \subset D$.

19. Show that a triple (X, Y, F) is a function if and only if:

    (a) (X, Y, F) is a correspondence;

    (b) F is a functional graph;

    (c) $pr_1(F) = X$.

20. Let G be a functional graph. Show that $(pr_1(G),\ pr_2(G),\ G)$ is a surjection.

21. Let G be a functional graph and E a set. Show that $G[G^{-1}[E]] \subset E$. Give an example of a graph R and a set F such that $R[R^{-1}[F]] \not\subset F$.

22. Let G be a functional graph, and E be a set. Show that:

(a) $G^{-1}[G[G^{-1}[E]]] = G^{-1}[E]$;

(b) $E \subset pr_1(G) \Rightarrow G[G^{-1}[G[E]]] = G[E]$.

Answers and Hints to Selected Exercises

Chapter 1

1. (a) true; (b) false; (c) true; (d) false; (e) false.

2. (a) false; (b) true; (d) true; (e) true; (f) true; (g) false.

3. (a) true; (b) true; (c) true; (d) true; (e) true; (f) true.

4. Suppose $X \subset Y$. Then $A \in \mathscr{P}(X) \Rightarrow A \subset X \Rightarrow A \subset Y \Rightarrow A \in \mathscr{P}(Y)$, hence $\mathscr{P}(X) \subset \mathscr{P}(Y)$. Conversely suppose $\mathscr{P}(X) \subset \mathscr{P}(Y)$. Then since $X \in \mathscr{P}(X)$ we deduce $X \in \mathscr{P}(Y)$, hence $X \subset Y$.

5. (b) $\mathscr{P}(\mathscr{P}(\mathscr{P}(\varnothing))) = \{\varnothing, \{\varnothing\}, \{\{\varnothing\}\}, \{\varnothing, \{\varnothing\}\}\}$.

Chapter 2

1. (a) $\{1, 2, 3, 4, 5\}$; (b) $\mathbf{Z}$; (d) $\{1, 2, 3\}$; (e) $\{1, 2\}$; (f) $\varnothing$; (g) A_3; (h) A_2.

2. Suppose $A \cup B = A \cap B$. Then $x \in A \Rightarrow x \in A \cup B \Rightarrow x \in A \cap B \Rightarrow x \in B$ hence $A \subset B$. Similarly we show that
$$B \subset A.$$
Thus $A = B$. If $A = B$, then $A \cup B = A \cap B$ obviously.

3. (b)
$$
\begin{aligned}
&((\mathbf{C}A) \cap (A \cup B)) \cup (A \cap B) \\
&= (((\mathbf{C}A) \cap A) \cup ((\mathbf{C}A)) \cap B)) \cup (A \cap B) \\
&= (\varnothing \cup ((\mathbf{C}A) \cap B)) \cup (A \cap B) \\
&= ((\mathbf{C}A) \cap B) \cup (A \cap B) \\
&= ((\mathbf{C}A) \cup A) \cap B = X \cap B = B.
\end{aligned}
$$

4. Suppose $(A \cap B) \cup C = A \cap (B \cup C)$. Then $x \in C \Rightarrow x \in (A \cap B) \cup C \Rightarrow x \in A \cap (B \cup C) \Rightarrow x \in A$, hence $C \subset A$. Conversely $C \subset A \Rightarrow A \cap C = C \Rightarrow A \cap (B \cup C) = (A \cap B) \cup (A \cap C) = (A \cap B) \cup C$.

6. $(E - G) \cap (F - G) = (E \cap \mathbf{C}G) \cap (F \cap \mathbf{C}G) = (E \cap F) \cap (\mathbf{C}G \cap \mathbf{C}G) = (E \cap F) \cap \mathbf{C}G = (E \cap F) - G.$

Chapter 3

1. $\mathscr{P}(A \times B) = \{\varnothing, \{(1, 2)\}, \{(1, 3)\}, A \times B\}.$
   $A \times \mathscr{P}(A \times B) = \{(1, \varnothing), (1, \{(1, 2)\}), (1, \{(1, 3)\}), (1, A \times B)\}.$

2. (b) $\{(2, 3)\} \cap \{(3, 2)\} = \varnothing$; (c) $(\mathbf{N \times Q}) \cap (\mathbf{Q \times Z}) = \mathbf{N \times Z}.$

5. (a) $R \circ S = \varnothing$, $S \circ R = \{(0, 2), (1, 3)\}$; (b) $R \circ S = \{(0, 2)\}$, $S \circ R = \{(0, 2)\}$; (c) $R \circ S = \{(1, 0)\}, S \circ R = \varnothing$; (d) $R \circ S = \{1, 2, 3\} \times \{1, 2, 3\}$, $S \circ R = \{(0, 0)\}.$

6. $(w, z) \in (T \circ S) \circ R \Rightarrow$ there is x in X such that $(w, x) \in R$ and $(x, z) \in T \circ S \Rightarrow$ there is x in X and y in Y such that $(w, x) \in R$, $(x, y) \in S$ and $(y, z) \in T \Rightarrow$ there is y in Y such that $(w, y) \in S \circ R$ and $(y, z) \in T \Rightarrow (w, z) \in T \circ (S \circ R)$, hence $(T \circ S \circ R \subset T \circ (S \circ R)$. Similarly $(T \circ S) \circ R \supset T \circ (S \circ R)$.

7. $z \in R^1[B] \iff$ there is w in B such that $(w, z) \in R^{-1} \iff$ there is w in B such that $(z, w) \in R \iff z \in \{x|$ there is y in B such that $(x, y) \in R\}$, hence $R^{-1}[B] = \{x|$ there is y in B such that $(x, y) \in R\}.$

Chapter 4

1. Yes, for example $(\{0, 1, 2, 3\}, \{1, 2, 3\}, \{(0, 1), (1, 2), (2, 3), (3, 1)\}).$

2. Yes, No.

4. No, No, No.

5. Assuming $a \neq b$ we have dom $e_a = \mathscr{F}(X, X) = \{f_1, f_2, f_3, f_4\}$ where
   $f_1 = (X, X, \{(a, a), (b, a)\}, f_2 = (X, X, \{(a, b), (b, b)\}),$
   $f_3 = (X, X, \{(a, a), (b, b)\})$ and $f_4 = (X, X, \{(a, b), (b, a)\}).$
   Thus $e_a(f_1) = f_1(a) = a, e_a(f_2) = f_2(a) = b, e_a(f_3) = f_3(a) = a$ and $e_a(f_4) = f_4(a) = b.$

6. $\{1, 2\}^{\{1\} \times \{1,2\}} = \mathscr{F}(\{(1, 1), (1, 2)\}, \{1, 2\}) = \{(\{(1, 1), (1, 2)\}, \{1, 2\}, G_i) \mid i \in \{1, 2, 3, 4\}\}$ where $G_1 = \{((1, 1), 1), ((1, 2), 1)\},$
   $G_2 = \{((1, 1), 2), ((1, 2), 2)\}, G_3 = \{((1, 1), 1), ((1, 2), 2)\}$ and
   $G_4 = \{((1, 1), 2), ((1, 2), 1)\}.$

7. Yes. $C = (A, B, \{((0, 0), 1), ((0, 1), 1), ((1, 0), 0), ((1, 1), 1)\}).$

Chapter 5

1. (a) $f(\{0, 3\}) = \{a, c\}$; (b) $f(\varnothing) = \varnothing$; (c) $f^{-1}(b) = \{2\}$;
   (d) $f^{-1}(\mathbf{C}\{a, c\}) = \{2\}$; (e) $f(\{1\} \cap \{2\}) = \varnothing$; (f) $f^{-1}(\varnothing - Y) = \varnothing$.

2. $(\mathscr{P}(X), \mathscr{P}(Y), \{(A, f(A)) \mid A \in \mathscr{P}(X)\})$.

3. (a) Suppose H is surjective. Let $s \in X$, $t \in X$, $s \neq t$. Then there is $A \subset Y$
   and $B \subset Y$ such that $H(A) = \{s\}$ and $H(B) = \{t\}$. Since H is a function
   (verify) and $\{s\} \neq \{t\}$ we have $A \neq B$. But since $A = \{f(s)\}$ and $B =$
   $\{f(t)\}$ (verify) we may conclude $f(s) \neq f(t)$. Now suppose f is injective, let
   $A \in \mathscr{P}(X)$. If we take $B = \{f(x) \mid x \in A\}$ we have $H(B) = f^{-1}\{f(x) \mid x \in$
   $A\} = A$. Since A was arbitrary we conclude H is surjective.

4. (b) $\sigma(\{1\}) \cup \sigma(\{2\}) = \{1, 2\}$.

Chapter 6

1. $g \circ f = (X, Z, \{(1, 4), (2, 4), (3, 4), (4, 3)\})$;
   $f \circ g = (Y, Y, \{(1, 2), (3, 1), (2,1), (4, 1)\})$.

2. Suppose h' is an extension of h. Since $h' \circ k$ and $l \circ h$ have the same domain
   and range we have to show that for each x in A $h' \circ k(x) = l \circ h(x)$. But
   $h' \circ k(x) = h'(k(x)) = h'(x)$ and $l \circ h(x) = l(h(x)) = h(x)$. Since $H' \supset H$
   we deduce $h'(x) = h(x)$, hence $h' \circ k(x) = l \circ h(x)$. Conversely, suppose
   $h' \circ k = l \circ h$. We must show $H' \supset H$. If $(x, y) \in H$ we have $x \in A$ and
   $y = h(x)$. Then $h'(x) = h'(k(x)) = h' \circ k(x) = l \circ h(x) = l(h(x)) = h(x) = y$,
   hence $(x, y) \in H' = \{(x, h'(x)) \mid x \in A'\} (l = j_{B,B'}$ and $k = j_{A,A'})$.

4. The definition is given in exercise 2 only for the case where f and g are
   functions. Since g is not a function we may not say that g is an extension of f.

6. (b) Let s and t be arbitrary elements of S. There are four cases according as
   $(\tau(s), \tau(t)) = (0, 0)$, $(0, 1)$, $(1, 0)$, or $(1, 1)$. In the chart below the values
   taken on by a given function in each case are written below the symbol
   representing that function:

$$M \circ (\tau \times \tau)(s, t)$$

0	0	0
1	0	1
1	1	0
1	1	1

$$\sigma \circ (\sigma \circ (\tau, \tau) \times \sigma \circ (\tau, \tau))(s, t)$$
$$= \sigma \circ (\sigma \circ (\tau, \tau)(s), \sigma \circ (\tau, \tau)(t))$$

0	1	0	0	1	0	0
1	1	0	0	0	1	1
1	0	1	1	1	0	0
1	0	1	1	0	1	1

Since the values given in the left-hand column are equal in each of the four cases we conclude that the two functions are equal.

Chapter 7

1. $\sigma^{-1} = (X, X, \{(2, 1), (1, 2), (3, 3)\})$.

2. $f^{-1} = (Y, X, G^{-1})$ (see Ex 7, Section 3).

3. If $n = 1$ the conclusion is obvious. Suppose $n > 1$. Then $f^{n-1} \circ f = f \circ f^{n-1} = j_X$ and we may apply Theorem 7.1.

Chapter 8

1. (a) $\{1, 2, 3, 4, 5, 6\}$;   (b) N;   (c) $\{2\}$;   (e) $\varnothing$;   (f) N;
    (g) $\{f_1, f_2, f_3, f_4\}$ where $\operatorname{dom} f_i = \{1, 2\}$ and $\operatorname{rng} f_i = \{1, 2, 3\}$ for all i in $\{1, 2, 3, 4\}$ and $f_1(1) = 1$, $f_1(2) = 2$; $f_2(1) = 1$, $f_2(2) = 3$; $f_3(1) = 2$, $f_3(2) = 2$; $f_4(1) = 2, f_4(2) = 3$; (h) A_8.

2. Denote the bijection by F with $\operatorname{dom} F = \{f_1, f_2, f_3, f_4\}$ (see Ex. 1) $\operatorname{rng} F = A_1 \times A_2$ and $G_F = \{(f_1, (1, 2)), (f_2, (1, 3)), (f_3, (2, 2)), (f_4, (2, 3))\}$.

3. Consult the definitions.

5. (a) $\{1, 4, 9, 16, 25, 36\}$;   (b) $\varnothing$;   (c) $\{1\}$;   (d) N;   (e) $\varnothing$;
    (f) $\{k^2, (k + 1)^2\}$.

6. $\bigcup \mathscr{G} = \{1, 2, 3, 4\}$, $\bigcap \mathscr{G} = \{3\}$ for the first case;
    $\bigcup \mathscr{G} = \{n \mid n \geq 2\}$, $\bigcap \mathscr{G} = \varnothing$ for the second case.

Chapter 9

1. Let u and v be elements of Y. If $u \neq v$ then $pr_k(u) \neq pr_k(v)$ for some $k, 1 \leq i \leq 5$. For this k we have $pr_k(u) = 1$ and $pr_k(v) = 0$, or $pr_k(u) = 0$ and $pr_k(v) = 1$. Thus $\{i \mid pr_i(u) = 1\} \neq \{i \mid pr_i(v) = 1\}$ and $\varphi(u) \neq \varphi(v)$. Therefore φ is injective. Now let $A \in \mathscr{P}(I)$. Define $(a_1, a_2, a_3, a_4, a_5)$ in Y by $a_i = 1$ if $i \in A$ and $a_i = 0$ if $i \notin A$ for each i in I. Then clearly $\varphi(a_1, a_2, a_3, a_4, a_5) = A$, hence φ is also a surjection.

Chapter 11

1. Since A is not empty we can pick $y \in A$. By the definition of inf and sup we have $\inf A \leq y$ and $y \leq \sup A$. Since $\leq$ is transitive we conclude $\inf A \leq \sup A$.

2. We show for example that R^{-1} is transitive: $(x, y) \in R^{-1}$ and $(y, z) \in R^{-1} \Rightarrow$ $(y, x) \in R$ and $(z, y) \in R \Rightarrow (z, x) \in R \Rightarrow (x, z) \in R^{-1}$. Let $A \subset X$, and suppose $\inf_{R^{-1}} A$ exists. Denote $\inf_{R^{-1}} A$ by a. Then $(a, x) \in R^{-1}$ for every x in A and if $(b, x) \in R^{-1}$ for every x in A we have $(b, a) \in R^{-1}$. This means that $(x, a) \in R$ for every x in A and if $(x, b) \in R$ for every x in A we have $(a, b) \in R$. By the definition of sup we conclude that $\sup_R A$ exists and $\sup_R A = a$. Similarly, on the assumption that $\sup_R A$ exists, one shows that $\inf_{R^{-1}} A$ exists and $\inf_{R^{-1}} A = \sup_R A$. The last statement may be deduced by taking $R = S^{-1}$ (S is an order relation) and using the above result.

Chapter 12

1. Let S be the set of all n in N such that

$$\sum_{i=1}^{n} i^2 = \frac{n(n + 1)(2n + 1)}{6}.$$

Clearly $1 \in S$. Suppose $k \in S$. Then

$$\sum_{i=1}^{k+1} i^2 = \left(\sum_{i=1}^{k} i^2 \right) + (k + 1)^2 = \frac{k(k + 1)(2k + 1)}{6} + (k + 1)^2$$

$$= \frac{1}{6} (k + 1)(k(2k + 1) + 6(k + 1)) = \frac{(k + 1)(k + 2)(2(k + 1) + 1)}{6}$$

hence $(k + 1) \in S$. By Theorem 12.1, $S = N$.

Chapter 13

1. (a) 10; (b) assuming $a \neq 3$, $a \neq \varnothing$, $c(J_X) = 3$; (c) 35; (d) 24; (e) 16; (f) 3.

3. Let $n \in N$ and $1 \leq p \leq n$. Then

$$\binom{n}{p - 1} + \binom{n}{p} = \frac{n!}{(p - 1)!\,(n + 1 - p)!} + \frac{n!}{p!\,(n - p)!}$$

$$= \frac{(n!)(p + (n + 1 - p))}{p!\,(n + 1 - p)!} = \binom{n + 1}{p}.$$

Bibliography

1. Bourbaki, N.: Théorie des Ensembles. Hermann (Paris), Chaps. 1–4, 1968.

2. Cohen, P. J.: Set Theory and the Continuum Hypothesis. W. A. Benjamin, Inc., 1966.

3. Cohen, P. J.: The independence of the continuum hypothesis, I, II. Proc. Nat. Acad. Sci., *50* (1963) and *51* (1964).

4. Cohen, L. W., and Ehrlich, E.: The Structure of the Real Number System. D. van Nostrand, 1963.

5. Halmos, P. R.: Naive Set Theory. D. van Nostrand, 1960.

6. Lang, S.: Analysis II. Addison-Wesley, 1969.

7. Morse, A. P.: A Theory of Sets. Academic Press, 1965.

8. Rudin, W.: Principles of Mathematical Analysis. McGraw-Hill, 1953.

9. Stone, M. H.: Number Systems of Analysis and Geometry. Holt, Rinehart and Winston (in preparation).*

10. Suppes, P.: Axiomatic Set Theory. D. van Nostrand, 1960.

---

*See also Stone, M. H.: Real number system reviewed. L'Enseignement Mathématique, *11*:263-267, 1969.

INDEX OF NOTATIONS

INDEX

DATE DUE